AUTHENTICITY AND WOODEN ARCHITECTURE PRESERVATION IN ASIA – A CHINESE PERSPECTIVE

The tradition of Chinese wooden architecture dates back to ancient times. The construction solutions developed in this country over the centuries enchant with their refined character, while the historical wooden structures delight future generations with their dignity and aesthetic excellence. China's wooden architecture, deeply rooted in its spiritual and religious traditions, is undoubtedly the pinnacle of this type of building in Asian culture. At the same time, it is a testimony to the national identity and reflects the specificity of the country's material heritage.

The accelerated social and economic changes in China and the constantly advancing globalization of the world have contributed to this country's assimilation of Western concepts related to the protection of cultural heritage. The issue of authenticity in the preservation of wooden built heritage proved particularly problematic. This book brings closer the theoretical understanding and practical application of the idea of authenticity from Chinese perspective. To do this, the issue of living heritage and the reception and understanding of traditional Chinese wooden architecture and its preservation as a direct materialization of Chinese religious and philosophical traditions is discussed. The above topics are treated within the cyclic concept of time, i.e. in terms of progress and repetition, with preservation being understood as a religious practice. Finally, trends in the preservation of wooden heritage in present-day China are mentioned, including new attempts to interpret the tradition and the reinvention of the tradition of wooden building.

The book aims to contribute to the understanding of the protection of wooden architectural heritage in China from a new perspective, and will be of particular interest to academics and professionals interested in or involved in the preservation of built wooden heritage.

'(…) a highly valuable contribution to the field of wooden architecture protection and preservation', Xiaoming Zhu Tongji University, Shanghai, China

This book '(…) successfully explains the inheritance characteristics of Chinese wooden architectures from the perspective of cultural philosophy for a wide audience (…)', Yasufumi Uekita, University of Tsukuba, Ibaraki, Japan

Authenticity and Wooden Architecture Preservation in Asia – a Chinese perspective

Tomasz Tomaszek

CRC Press
Taylor & Francis Group
Boca Raton London New York

CRC Press is an imprint of the
Taylor & Francis Group, an **informa** business
A BALKEMA BOOK

Designed cover image: Tomasz Tomaszek

First published 2023
by CRC Press/Balkema
4 Park Square, Milton Park, Abingdon, Oxon, OX14 4RN
e-mail: enquiries@taylorandfrancis.com
www.routledge.com – www.taylorandfrancis.com

CRC Press/Balkema is an imprint of the Taylor & Francis Group, an informa business

Library of Congress Cataloging-in-Publication Data

A catalog record has been requested for this book

ISBN: 978-1-032-33179-9 (hbk)
ISBN: 978-1-032-33181-2 (pbk)
ISBN: 978-1-003-31859-0 (ebk)
DOI: 10.1201/9781003318590

Table of contents

Prior to acceptance, the final manuscript has been reviewed by Professor Xiaoming Zhu, of the College of Architecture and Urban Planning, Tongji University, Shanghai, China, and by Professor Yasufumi Uekita, of the Heritage Studies Program, Faculty of Art and Design, University of Tsukuba, Tsukuba, Japan.

Preface

The term "authenticity", according to its Greco–Roman origins, can be translated directly as "genuine" and "true". After it was introduced for the first time in an international conservation document called the Venice Charter (ICOMOS, 1965), authenticity become one of the most discussed issues in monument conservation. However, as the principles of the Venice Charter were formulated initially within a Western European conceptual and ideological framework, the applicability of these principles together with the unequivocal notion of authenticity quickly became questioned in the context of various cultures. In particular, the notion of authenticity became ambiguous when applied in the context of non-European traditions. Additionally, the mostly archaeological approach to the monuments focusing on the original fabric, which directly resulted from the principles proposed by the Venice Charter, was particularly difficult to reconcile in the case of wooden buildings, especially in Asian countries. Therefore, the term of authenticity faced the necessity of extended understanding of its notion, which was later underlined in *The Nara Document on Authenticity* announced in the year of 1994 (ICOMOS, 1994).

Among the Asian countries with the richest tradition of wooden architecture China stands out. This is where some of the most magnificent and oldest wooden buildings in the world are located. At the same time, China can boast countless historically developed technological solutions in terms of the construction of wooden buildings, which reflect the extremely rich spiritual and religious traditions of this country. Thus, wooden architecture, being the core of China's material culture/tangible heritage, at the same time perfectly illustrates the intangible aspects of which it is materialization and substantiation.

What also need to be stressed, is that Chinese wooden architecture for centuries was created under enormous influences of traditional Chinese philosophy, thus the usage of wood and wood as the fabric brings layers of various meanings to the architectural structure. However, at the same time, it must be remembered that due to all these aspects mentioned above the traditional wooden buildings in China were more of an exhibition of social status or religious meaning, whereas the fabric and structures involved were not taken seriously as a technology *per se* for a long time (Ze-li, et al., 2017, p. 198) and could be treated as the "specific carriers of cultural memory".

It is wood which has an essential influence on the joints and the performance of the whole structure, which makes Chinese architecture so unique. But also, parallel, because of properties of the wood, Chinese traditional wooden structure is characterized by lack of durability and need for regular maintenance and renovation that guarantees the continuity of the architectural intention and spiritual purpose. As a consequence, the periodical reconstruction of wooden structures were practiced in China for centuries, which resulted in developing of some unique approaches to so called conservation practice.

The above mentioned factors played a fundamental role at the time when China opened up to the influence of international conservation principles and ideas, and thus adopted in consequence the term of authenticity in heritage conservation. The strong influence of international legislation and Western conservation philosophies clashed with the extremely distinct tradition of Chinese culture, shaped by millennia. The rich heritage of wooden architecture and its centuries-old conservation traditions has taken over the Western concept of authenticity, giving it its own meaning and understanding, and placing it in the conservation practice of contemporary China in a very specific way. The reading of these given meanings and the references to authenticity in the conservation of wooden architecture in China is as fascinating as the sheer power and finesse of traditional Chinese wooden building.

This monograph is an attempt made by a researcher belonging to the Western culture to face the problem of understanding and applying the concept of authenticity in the context of Chinese culture. In a way, this intention directly reflects the very phenomenon of the absorption and settlement of the Western term of authenticity together with its Western notion in the realities of a different culture. The author, coming from the Western world, while facing a cultural reality very different from his own, has the chance to observe how the term he knows from Western philosophies of conservation has been adapted to the context of Chinese culture. However, the task undertaken by the author seems justified, because it is thanks to this extraordinary situation that the Western researcher analyzes the term directly derived from his own culture, at the moment when this term is transferred to a new cultural ground. Despite the difficulties of the project, the task undertaken seems to be extremely valuable, as it initiates a discourse on the issue of authenticity in the protection and conservation of wooden architecture in China, and thus contributes an important voice to the discussion on the protection of the heritage of wooden architecture in general.

This monograph is also an attempt to deal with this specific "tension" noticeable in today's China, and resulting from a direct clash of two opposing forces – on the one hand, the still exceptionally strong traditional Chinese culture materializing its intellectual and spiritual achievements in wooden architecture, and on the other hand, an increasing implementation of a whole variety of Western culture patterns into traditional Chinese lifestyle. These patterns are noticeable both in the everyday reality of China and in the trends and practice of interpreting and protecting cultural heritage. The conflict of these two worlds can be seen, among other things, in the dynamics of understanding and applying the concept of authenticity in China. Thus there are two different attitudes facing each other – shaped in the West, the belief in the extraordinary value of remnants of historical fabric and a one-time expression of design, versus the recognition of intangible factors as the kind of essential determinant of the originality and truthfulness of a historical architectural structure found in Eastern cultures. The collision of these two separate worlds results not only in a clash between the opposites, followed by the dominance of one of two mentioned notions of authenticity, but above all in giving new levels of interpretation and meaning to this value, as well as in a particular application of this concept in conservation practice. All this illustrates today's China – a country of enormous contrasts, where the core of Asian culture is often dressed with contrariness in Western garments. It is a task for itself in this study to read this particular tension mentioned above and its impact on the wooden built heritage conservation model that is currently taking shape in China.

In the presented monograph, a special place is also occupied by considerations of authenticity in the context of living heritage. This aspect seems to be particularly relevant to China. This country, functioning in the model of the communist organization of society, went through a significant for its history period of so-called "cultural revolution", when all tangible and intangible manifestations of "backward" tradition and spirituality were consistently liquidated. The aspirations were to lead to the development of a uniform society and modern culture stripped of the remnants of previous habits and inheritance. Surprisingly, this intention did not work. On the contrary, China is still a country with an extremely strong local tradition and ubiquitous living heritage. Worth mentioning here are, among others, the mountainous areas inhabited by ethnic minorities (for example the Yunnan province where wooden vernacular log building culture is still very alive) or the Tibet Autonomous Region. It is in here that the living heritage manifests itself in a very special way, constituting the core of this place and its above-average attractiveness for Western tourists. It can even be said that Tibet is one of the last places where living heritage manifests itself in its purest form. Continuation of the tradition of wooden architecture in China, from the perspective of the issue of living heritage, seems essential for a proper understanding of the notion of authenticity. It is also worth emphasizing that China, with all its aspirations to become a country open to other cultures, still remains one of the most closed communities in the world. This directly influences the significance of living heritage in China as the guarantor of the transmission of local traditions and the maintenance of the core of national identity. Here, too, it can be noticed this specific "tension" found in today's China – on the one hand, an ultra-modern country, with advanced technology and adapted patterns of Western culture, and on the other, very conservative and traditional, where established patterns of cultural behavior are inviolable matrices.

The monograph is complemented by a brief description of the development and main construction types of Chinese wooden architecture, as well as examples of the most significant historic wooden buildings in China, the oldest of which remember the distant times of the Tang Dynasty.

It remains problematic to discuss the issue of authenticity, in the context of Chinese culture, in relation to Tibet. While maintaining its autonomy, Tibet is viewed by many as an entirely separate state and an independent country. Leaving this difficult issue to others to resolve, Tibet in this monograph is treated as a certain element of the overall organism of China, understood not as a country covered by specific borders, but more a cultural area governed by a similar or even an identical understanding of the principles of the assumed spiritual or philosophical reality. At the same time, it is worth noting that, like many other cultural regions in the world, China does not constitute a uniform system without separate subsets. And looking from this perspective, Tibet naturally fits into this reality, as a component of an extremely distinct and interrelated cultural area on the Asian continent.

The research results presented in the monograph, are the outcome of several study visits/research stays by the author, and the author's research projects carried out at the universities and research institutes/centers in China, as well as author's extensive field studies carried out in Mainland China and Tibet Autonomous Region. And so the following should be mentioned:

- Participation in projects on protection, conservation and monitoring of historic wooden architecture in China, as a representative of Poland in the Course on Management and Monitoring of World Heritage Sites with special reference to China, which took place on the site of the Summer Palace World Heritage Site, Beijing, in 2016. The project was organized by SACH – State Administration of Cultural Heritage in China; CACH – Chinese Academy of Cultural Heritage and ICCROM – International Centre for the Study of the Preservation and Restoration of Cultural Property.

- The implementation of a scientific fellowship and a research project at the Institute of Architecture at Chang'An University in Xi'An in 2017-2018, thanks to funding (Senior Scholar Fellowship) obtained from the government of the People's Republic of China (Ministry of Education of the People's Republic of China) under the CSC program (China Scholarship Council). The research was carried out with the participation of researchers of the Institute of Architecture of Chang'An University, inter alia Prof. Liu Wei. A significant part of the research was carried out in situ, on the most important objects of the historical wooden architecture of the city of Xi'An, the first capital city of China. The research was carried out, among others, on the Bell Tower, the Drum Tower, some parts of the ancient city walls and selected Buddhist temples. In addition, research was carried out on significant historical wooden buildings located in various provinces of the People's Republic of China (including Guangxi Province – on wooden "drum towers" built by the ethnic minority of the Dong people; and Sichuan Province – on wooden Buddhist temples located in Emei Shan massif, which is one of the "Sacred Mountains of Buddhism" [the whole area is inscribed on the UNESCO World Heritage List]. At the same time, an important part of the research was carried out in the Chinese capital, Beijing (where field studies were carried out on wooden architectural structures within the "Forbidden City" – the Palace of Chinese Emperors, and on the wooden buildings of the Summer Imperial Palace). In addition, part of the field studies were carried out in the Tibet Autonomous Region (including the architectural complex of the Potala Palace and selected Tibetan monasteries in the Himalayan foothills, including monasteries located at the foot of Mount Kailash).

- The implementation of a research fellowship as a Senior Scholar and a research project at AATU – Tianjin University Research Institute of Architectural Design and Urban Planning, Tianjin University, Tianjin, and the School of Architecture at the Tianjin University in Tianjin during the years 2019-2020. The research was possible thanks to a research fellowship/research grant obtained under "The Great Wall" program, financed by UNESCO in cooperation with the government of the People's Republic of China (Ministry of Education of the People's Republic of China/China Scholarship Council). Obtaining the abovementioned research grant was possible thanks to the support of The Polish National Commission for UNESCO (support of the candidacy as the only one representative from Poland for this program due to research achievements, confirmed by a letter of recommendation issued by the Commission's Secretary-General, Prof. Sławomir Ratajski). The research was carried out with the participation/cooperation of researchers of the AATU Research Institute, particularly Prof. Zhang Wei, a recognized architect responsible for the protection and conservation of the historic wooden architecture of the direct-administrated municipality of Tianjin, including one of the oldest wooden buildings in China - the Buddhist Dule Temple. A considerable part of the research, was carried out in situ on selected objects of China's wooden architectural heritage. In addition to the studies carried out at the above-mentioned Dule Temple, a significant part of the scientific research was carried out in the area of Wutai Shan – a group of five mountain peaks constituting the area called "Mount Wutai", one of the most important Sacred Mountains of Buddhism and the place with the greatest concentration (largest number) of historical wooden Buddhist temples in China. Among the objects covered by the research are the two oldest wooden temples in China (also the oldest wooden buildings in China) – the Nanchan Temple (built during the Tang Dynasty in 782) and the Foguang Temple (built in 857).

Courtesy of the Faculty of Architecture at Kunming University in Kunming, Yunnan Province, part of the research was carried out in situ in the Yunnan Province in collaboration with Prof. Yang Dayu, a valued expert in vernacular architecture in the southern part of China. The scientific research focused mainly on the historical log architecture present in these areas and covered the area around Lake Lugu (one of the largest concentrations of log architecture in China) and the historic wooden settlements of ethnic minorities living in the Yunnan province (including the traditional village of Tongle, located in hardly accessible mountain areas and entirely consisting of historic log buildings, and which village is typical example of living heritage in China).

Acknowledgements

The author would like to express a special thanks to Professor Zhang Wei, of Tianjin University Research Institute of Architectural Design and Urban Planning/the School of Architecture at the Tianjin University in Tianjin; and Professor Yang Dayu, of the Faculty of Architecture at Kunming University in Kunming, Yunnan Province for their help and commitment in carrying out this research, the summary of which is this monograph.

About the author

Dr Tomasz Tomaszek works as an Associate Professor at the Department of Monuments Conservation, Faculty of Civil and Environmental Engineering and Architecture at Rzeszow University of Technology in Poland. He graduated with an MFA in Conservation and Restoration of Fine Art Objects from the Academy of Fine Arts, Krakow, Poland and an MA in Philosophy, Jagiellonian University, Krakow, Poland. He obtained his PhD (with honors) in Technical Science in Discipline of Architecture and Urbanism (specialty: history and conservation of monuments of architecture and urbanism) at Krakow University of Technology, Faculty of Architecture, Department of the History of Architecture and Conservation of Monuments, Krakow, Poland.

Dr Tomaszek furthered his professional qualifications by participating in several international specialist courses and additional studies, including:

- International Course on Wooden architecture conservation and restoration, organized by ICCROM, Petrozavodsk State University, UNESCO Chair "Wooden Architecture Research and Preservation" and Ministry of Culture of the Russian Federation; Petrozavodsk and open-air museum Kizhi, Russia;
- Course on Management and Monitoring of World Heritage Sites with special reference to China, organized by State Administration of Cultural Heritage China, ICCROM and China Academy of Cultural Heritage; The Summer Palace World Heritage Site, Beijing, China;
- International Course on Conservation of Built Heritage (CBH14), organized by ICCROM, Rome, Italy;
- 14th International Course on Wood Conservation Technology (ICWCT 2010) – A Course on the Conservation of Cultural Heritage Made of Wood, organized by Riksantikvaren - The Directorate for Cultural Heritage in Norway, Oslo, Norway; and "Architecture – Continuing Education", Norwegian University of Science and Technology, Trondheim, Norway

Dr Tomaszek is a laureate of many prestigious scholarships/fellowships, such as:

- Senior Scholar Fellowship at AATU – Tianjin University Research Institute of Architectural Design and Urban Planning, Tianjin, China; research fellowship from UNESCO/People's Republic of China (The Great Wall) Co-Sponsored Fellowships Programme
- Senior Scholar Fellowship at Institute of Architecture, Chang'An University, Xi'An, China; research fellowship from China Scholarship Council;
- Research Fellowship at the Center for Historic Preservation, Middle Tennessee State University, Murfreesboro, Tennessee, USA; research grant from the Kosciuszko Foundation;
- Study visit in Riksantikvaren - The Directorate for Cultural Heritage in Norway, Oslo, Norway – study visit was possible due a scholarship/grant from FSS Scholarship and Training Fund;
- Scholarships (three times) "Polish Culture Worldwide Programme", Adam Mickiewicz Institute, Poland.

Dr Tomaszek has presented his research at many national and international scientific conferences (inter alia in Palestine, the Czech Republic, Egypt, Lithuania, Portugal and Italy), and he participated in an international research grant (Japanese-Ukrainian-Polish co-operation) "Conservation Method of Wooden Churches in Ukraine – Comparative Study of

Log Construction Conservation Method between Japan and Europe"; research fund from Japan Society for the Promotion of Science (JSPS).

Dr Tomaszek is the author of several scientific papers on issues related to the preservation of historical wooden architecture, which have been published in renowned Polish and international journals.

A brief introduction to traditional Chinese wooden architecture

ABSTRACT: Traditional Chinese wooden architecture is one of the world's three major architecture systems (besides European and Islamic architecture) and without doubt plays an essential role in the global history of architecture. What is particularly worth underlining, Chinese architecture developed a particular system of construction that has retained its principal characteristics from ancient times to the present day. This chapter presents a brief history and characterizes the main types of wooden building in China, including the vernacular building tradition of ethnic minorities. It is also discusses the Chinese understanding of space and layout of architectural design, as well as the philosophical and spiritual aspects that have influenced Chinese building tradition for millennia. Additionally, architectural painted decoration is presented along with its traditional use and levels of meaning. Photographs provide illustrations of the historic buildings, showing the richness of Chinese wooden built heritage and its outstanding beauty.

Traditional Chinese architecture without a doubt plays an essential role in the global history of architecture. During its very long history, it has developed unique systematic features that testify to its uniqueness. It needs to be stressed that besides European and Islamic architecture, it is one of the world's three major architecture systems. Therefore, its esthetic and design solutions have had an influence on almost every corner of the globe. Additionally, it is in China where abundant wooden architectural heritage has survived, which shows thriving local building development through the millennia (Figure 1.1 – 1.10).

As noted by Gao, Chinese architecture developed "an indigenous system of construction that has retained its principal characteristics from prehistoric times to the present day" (Gao, 1984, p. 66). According to historians, its tradition emerged from a system using earth and wood, changing with time to one using bricks and wood. After thousands of years of progression and evolution, Chinese architecture has formed a complete system of structure and construction, which includes building regulations and standards inherited both from Song Dynasty (1103 AD) and from Qing Dynasty (1734 AD) (Ze-li, et al., 2017, p. 197).

However, apart from the main styles so-called "traditional Chinese architecture", various types of "local building tradition" have been developed by the Chinese. In the majority of cases, these building traditions belong to the ethnic minorities and can be find in almost all provinces of continental China as well as Tibet Autonomous Region (Figure 1.11 – 1.14). This historic vernacular architectural heritage of different regions. It uses unique construction methods as a result of different geographical characteristics as well as different cultural backgrounds, status or affiliations (Jin, 2019, p. 81). What is particularly worth mentioning is vernacular log architecture is abundant in north-eastern China or Yunnan Province. Much as it is, to a local extent, the execution of the construction system that is also typical for other regions of the World. Namely Russia or Eastern Europe. Thus, the wealth of local architectural solutions shows the enormity of achievements of the building tradition in China, and wooden structures occupy a leading position. Interestingly, the historical architecture of China is very often identified mainly with typical, classic solutions known from the significant monuments of this country. It is the achievements of vernacular architecture that fully reflect the character of this country's built heritage. It is also worth emphasizing that the above-mentioned log construction solutions, for example from the territory of the Yunnan province, prove similarities in the development of formal solutions

Figure 1.1. The Hall of Supreme Harmony (TaiHe Dian). Commonly known as the Hall of Golden Bells (Jinluan Dian), the Forbidden City, Beijing. It was built in the Ming dynasty in 1420. It is the highest-ranking ancient building still in existence in China, and accordingly the eaves feature the highest level of decoration. The hall has been burned down and been rebuilt several times. The structure we see today was rebuilt in 1695 (photo T. Tomaszek, 2018).

Figure 1.2. The Hall of Supreme Harmony (TaiHe Dian). Commonly known as the Hall of Golden Bells (Jinluan Dian), the Forbidden City, Beijing; detail (photo T. Tomaszek, 2018).

Figure 1.3. The Gate of Supreme Harmony, the Forbidden City, Beijing. It was originally built during the Ming Dynasty, it was destroyed 1886 by fire started by a tipped lamp in the guard room. The present gate dates from the rebuilding after this fire, which was completed in 1894 (photo T. Tomaszek, 2018).

Figure 1.4. The Gate of Supreme Harmony, the Forbidden City, Beijing; interior (photo T. Tomaszek, 2018).

resulting from the limitations of the material (in this case wood in the form of log), regardless of the different cultural regions in which they are used. The log construction in Yunnan shows a specific similarity to such log buildings in Russia, Europe or other regions, while being historically modified in relation to the geographical and cultural context of China. And this, in turn, makes us aware of the extremely complex patterns of intercultural influences in the past centuries and, in contrast, some observable formal and construction tendencies in architecture that are supra-cultural in nature.

Figure 1.5. The Gate of Heavenly Purity, also named Gate of Celestial Purity or Qianqingmen, the main gate of the Inner Court of the Forbidden City, Beijing. It was established in 1420 and rebuilt in 1655 (photo T. Tomaszek, 2018).

According to Hao, the history of ancient Chinese architecture can be divided into three basic periods. These are the time of primitive society, then slave society and finally feudal society. The latter period, is the time when Chinese classical architecture developed the most (Hao, 2004, p. 104). The beginnings of Chinese architecture should be sought about 7000 years ago. It dates back to the period of primitive society, which in architecture it lasts to the 21st century BC. During this historic time, different types of buildings were erected. They vary due to different climates, geographical features and materials in various parts of contemporary China. Among architectural structures two types were particularly distinguished: wooden frame and mud wall buildings, which most likely evolved from cave houses popular in the Yellow River basin, and Ganlan-style buildings (wooden buildings built on stilts) developed most likely from nest houses typical for Yangtze River basin. In the late period of primitive society, the walls and roofs of buildings were mostly made of interwoven branches or twigs covered with mud (Hao, 2004, p. 106). With the arrival of the 21st century BC, the wooden frame and rammed earth construction prevailed. As well as groups of buildings with an enclosed courtyard, which were already marked by a significant improvement in the technology of the wooden frame. Thus beginning of the next historic period of the development of Chinese architecture is dated from this point. It is called the period of slave society.

Figure 1.6. The Gate of Compassion and Tranquility (Cining men), the main entrance to the Palace of Compassion and Tranquility (Cining gong) in the western sector of the Inner Court, Forbidden City, Beijing. It was constructed during the Qianlong reign (1736–1795) of the Qing dynasty (photo T. Tomaszek, 2018).

Figure 1.7. The Gate of Compassion and Tranquility (Cining men), the main entrance to the Palace of Compassion and Tranquility (Cining gong) in the western sector of the Inner Court, Forbidden City, Beijing; interior (photo T. Tomaszek, 2018).

Figure 1.8. Wanfuge Pavilion (built 1748-1750) in the Lama Temple (Yonghegong), the biggest Tibetan Buddhist temple in Beijing, built in 1694 as the residence of Prince Yong of the Qing Dynasty, then in 1744 converted into a temple. In 1961, it was listed as a major historic site under state protection. The Yonghegong Lama Temple survived the ten turbulent years of Cultural Revolution from 1966 to 1976, thanks to Premier Zhou Enlai (photo T. Tomaszek, 2018).

Figure 1.9. Interior of Falun Hall (built in 1744) in the Lama Temple (Yonghegong), the biggest Tibetan Buddhist temple in Beijing, built in 1694 as the residence of Prince Yong of the Qing Dynasty, then in 1744 converted into a temple (photo T. Tomaszek, 2018).

Figure 1.10. The Bell Tower in Xi`an, Shaanxi Province, built in 1384. It is an example of the typical architectural style of Ming Dynasty. It has undergone many repairs since 1949. Now it is an important historical unit under national government protection (photo T. Tomaszek, 2018).

Figure 1.11. Entry Gates at Zhaoxing Dong Village, Guizhou Province. The village, belonging to the ethnic minority of Dong, was built in 1160 (in the 5th year of the Southern Song Dynasty). It is the most famous of the group of eight Dong Villages, called sometimes by similar name as "Zhaoxing Dong Villages". All of these villages are famous for their wooden drum towers and they are rated as "one of the six most beautiful ancient villages and towns in China" (photo T. Tomaszek, 2018).

Figure 1.12. Xintuan Drum Tower, at Zhaoxing Dong Village, Guizhou Province. In Dong language called Lou Dou Pe, belongs to the clans of Bai and Ma. It was first built in the 18th century and rebuilt in 1982, designated by the carpenter Lu Jixian from Tang`An (photo T. Tomaszek, 2018).

Figure 1.13. Traditional log houses on a hillside of the remote Tongle village (near Yezhi), Yunnan Province (photo T. Tomaszek, 2019).

Figure 1.14. A traditional log house in the remote Tongle village (near Yezhi), Yunnan Province (photo T. Tomaszek, 2019).

According to Hao, it lasted from 2070 until 476 BC. (Hao, 2004, p. 104). The peak of the development of the Chinese slave society fell in the 16th century BC, the time when of the beginning of documentary trace began. Based on the size of the rammed earth foundation of the palaces and temples, it should be assumed that the buildings in those times had a larger scale and a more strict hierarchy. Therefore, the scale of cities, city walls heights, street`s widths or the size of other significant buildings had to be built according to their rank. In the Spring and Autumn periods (770–476 BC), the most important achievements were the popularization of tiles and the appearance of buildings erected on high platforms (this concerned mainly imperial and ducal palaces). Building with a high-platform means building it with a platform of compacted earth underneath. As the leuds competed in building more magnificent palaces, the decoration and painting of ancient architecture was then mastered significantly (Hao, 2004, p. 108).

After the collapse of the slavery system, agriculture and handicrafts developed rapidly. In addition, iron products began to be used on a larger scale, which significantly improved construction methods and increased the quality of wooden structure`s construction. As noted by Hao, this brought around 475 BC beginning of the next historic period, called the period of feudal society, which lasted until 1911 BC (Hao, 2004, p. 106). During this period, fireplaces, heated brick beds and cellars became popular. A special heyday for classical Chinese architecture was the time of the Han Dynasty, when the commonly seen nowadays, the beam-lifted frame and through-type frame wooden structures were introduced. Also during that time, the traditional roof of Chinese buildings started to flourish. In turn, the introduction of Buddhism led to the rapid development of Buddhist architecture, one of the most important types of classical Chinese architecture (Hao, 2004, p. 108). During the next decades, during the rules of Tang Dynasty, the techniques and artistic qualities of classic architecture developed the fastest. The features of Tang-style architecture have become recognized for the extremity of dimensions and regulations, the extremity in a complex architectural layout, as well as large expansion and large volume. At that time, the construction form and material requirements for wooden structures, especially popular Dou-gong brackets, were also standardized. The Tang-style architecture that was flourishing in China at the time, had a far-reaching influence on Asian countries, especially Japan. Subsequently, during the time of Song Dynasty, the modular system was introduced and building standards were officially published, which let to establishing the standardized rules for buildings` measurements and basic moduli. This allowed the size of the wooden components to be properly defined (Ze-li, et al., 2017, p. 200).

In the late stages of feudal society times, the architectural structures took on an increasingly simplified form. Whereas the beam-and-column frame become one entirety. The buildings were erected according to stricter rules and the architectural decoration and painting were combined with the structure in a more sophisticated way. During the rules of Qing Dynasty, the ethnic diversity of the society influenced the emergence of a multitude of different residential types of housing. At that time, a certain homogeneous building form of official architecture was established. As a result, improved the standard of designing more advanced architectural complexes. The published construction rules listed 27 practices for unified building form and formulated new construction modules, which greatly contributed to accelerating the design and construction process, as well as for better control of material consumption (Pan, 2015).

Figure 1.15. The Prayer Hall of the Great Mosque in Xi`An, Shaanxi Province. The Mosque was built during the Tang Dynasty in 742 A.D. Then it was rebuilt and expanded under the Song, Yuan, Ming and Qing Dynasties. It was announced as a provincial key cultural relic protected unit in 1956 and was included in the third batch of the national key cultural units in 1988 (photo T. Tomaszek, 2018).

The majority of traditional Chinese architecture is constructed of wood and the main technology of its erection is carpentry (Figure 1.15 – 1.19). The usage of wood is not accidental. In ancient China, it was adopted as a primary building material not only because of its availability and simplicity of basic construction, but also because Chinese people believe that "life is connected with nature and humans should interact with animated things." (Li, et al., 2017, p. 132). Therefore, as the substance coming directly from the alive body of the tree, beam-column structure was very commonly adopted and appreciated in ancient Chinese architecture.

Figure 1.16. Interior of the Prayer Hall of Great Mosque in Xi`An, Shaanxi Province (photo T. Tomaszek, 2018).

Figure 1.17. Mahavira Hall of the Wenshu Baddish Temple (or Wenshu Monastery) located in downtown Chengdu City, Sichuan Province. It's the largest and the best-preserved Buddhist Monastery in Chengdu. The Temple was first built in the Sui Dynasty, between 605 - 617, and it was first called Xinxiang Temple. It was destroyed in Song Dynasty, then it was reconstructed in 1706 (photo T. Tomaszek, 2018).

Figure 1.18. Shakyamuni Temple at Kumbum Monastery (Ta`er Si Temple), Xining, Qinghai Province. It is one of the six major monasteries of the Gelug School of Tibetan Buddhism. The majority of the temples from the monastery were constructed after 1560. On March 4, 1961 The Kumbum Monastery was listed into the first batch of historical and cultural relics under state protection by the State Council (photo T. Tomaszek, 2018).

Figure 1.19. Bi Yong Hall, part of Guo Zi Jian (The Imperial College) at the Confucian Temple in Beijing (The temple, called KongMiao, was first built in 1302-1306, it is one of the Key National Culture Relic Protection Sites). Bi Yong Hall was built in 1783 and it once served as a venue for the emperor to conduct lectures. It is the only site of ancient institution of the highest education that has been preserved intact in China (photo T. Tomaszek, 2018).

With time two main wood structural styles have been developed. The Tailiang type (抬梁式) in the north and the Chuandou type (穿斗式) in the south. Traditionally, the wood components in typical Chinese architecture are connected by semirigid structural joints Sunmao (榫卯) and unique support Dougong (斗拱). These kind of joints and connection make the buildings flexible and easy to disassemble for repair and replacement (Jin, 2019, p. 84); (Figure 1.20 – 1.23).

Figure 1.20. The Main Hall of Confucian Temple (Da Cheng Dian, or Great Achievement Hall) in Pingyao, Shanxi Province. It dates to 1163 AD, in the Jin dynasty. The hall is 5 bays square, with a hip-gable roof and prominent eave brackets (dougong); (photo T. Tomaszek, 2018).

Traditional Chinese architecture, which is the best known for its official and high range structures as the temples or palaces, is famous for its symmetrical layout and directional hierarchies (Figure 1.24). Usually they [temples or palaces] consist the groups of buildings connected together following strict rules and commonly located along the central courtyard[1]. The latter is an inward enclosed open space which follows an axis symmetrical principle. Typically, all the most important buildings from these complexes are located facing South, with the two sides facing East and West respectively, and strictly aligned along a north/south axis (Kohrman, 1998, p. 236). This very particular order of alignment of individual buildings follows and represents a strong traditional hierarchical meaning. And so, beginning from the entrance of the temple of palace complex, the hierarchical space changes from open to closed, from public to private, from low to high rank. Most ancient Chinese buildings follow layouts "orienting the structure with its back to an elevated landscape and ensuring that there is water at the front" (Kohrman, 1998, p. 236). Traditionally, it was done in this way to avoid adverse geographical and weather

1. Complexes of buildings situated along the central courtyard and following strict rules of location between them are typical also for the historical type of the residences. Siheyuan (四合院) in Northern China and Tianjing (天井) in Southern China are typical examples.

Figure 1.21. The Main Hall of Confucian Temple (Da Cheng Dian, or Great Achievement Hall) in Pingyao, Shanxi Province. It dates to 1163 AD, in the Jin dynasty. Detail of eave brackets (dougong); (photo T. Tomaszek, 2018).

Figure 1.22. The city tower (Shi Lou), situated in the center of Pingyao Ancient City, Shanxi Province. It was built in ancient times (unknown date), and then it was rebuilt in 1688 (Qing Dynasty). The roof style was named "Xie Shang Ding" and decorated with colorful glaze (yellow and green); (photo T. Tomaszek, 2018).

Figure 1.23. The city tower (Shi Lou), situated in the center of Pingyao Ancient City, Shanxi Province. It was built in ancient times (unknown date), and then it was rebuilt in 1688 (Qing Dynasty); detail (photo T. Tomaszek, 2018).

conditions, and at the same time seek methods to integrate buildings into nature (Jin, 2019, p. 85). Since ancient times, Chinese architecture has used concepts of Feng Shui[2] to organize layouts and structures from common residences to Imperial and Religious structures (Knapp, 2006). It is also characterized by significant symbolism, which is mostly affected by the traditional Chinese concepts such as Confucianism, Taosim and Buddhism (Jin, 2019, p. 83).

Thus it is worth at this point to postulate, especially bearing in mind the Chinese understanding of space and layout of architectural design. That the landscape together with heritage resources (among others traditional wooden architecture) located on it, should be viewed as part of the cosmology of past and present societies. And that they together, as a complex system, form a platform for human-nature-environment interaction (Andrews and Buggey, 2008). However, very often, based on western philosophies concerning the conservation and management of heritage and particularly the internationally recognized expressions of authenticity and integrity (as by UNESCO through the 1972 World Heritage Convention (UNESCO, 1972), many heritage practitioners unfortunately look at Chinese architectural heritage basically through its physical attributes and as reflecting only past expressions and finalized materializations of chosen building tradition.

To gain a better understanding of historic Chinese architecture it is necessary to take a look at particular details of its design. And so, traditionally it possesses a horizontally developed lowering facade. Apart from some special functional structures, most classical wooden Chinese structures typically have one or two floors. It is due to the particular

2. Feng Shui (wind-water in English) was widely adopted to orient significant structures such as tombs, dwellings and other buildings in ancient China.

Figure 1.24. View over the roofs of groups of temple halls towards Kunming Lake, from the Temple of Buddhist Virtue, the highest point in Summer Palace, Beijing. The symmetrical layout and directional hierarchies of this imperial palace complex can be seen (photo T. Tomaszek, 2018).

Chinese philosophy of integration of the building into nature, as well as the limitations of wood construction. Therefore, the facade of the buildings stresses the visual impact of the width instead of height, as the buildings that were too high were considered unsightly (Yih-yuan, 1979, p. 332). As noted by Jin, even in the Forbidden City, the most significant Imperial architecture complex in China, with the highest hierarchy possible, the Halls and Palaces have rather low ceilings, which is in contrary with equivalent western buildings with similar hierarchy (Jin, 2019, p. 85).

Apart from common features such as the enclosed courtyard space, the hierarchies system, an axis symmetry or horizontal emphasis. Traditional Chinese architecture is characterized by outstanding architectural painting (Figure 1.25 – 1.37). Commonly important ancient Chinese buildings have brilliant colors and decorations consisting of traditional images and sculptures. The type of decoration, as well as the colors used, were chosen strictly according to the grade of the building, the location of the component and their material properties. Therefore, the decoration is not only to beautify the building, but also clarify its function and hierarchy. The particular colors, sculptures and other decorations combine with the whole building and its environment. Thus, underlying this very unique artistic realm and acquiring the special spiritual meaning through the narrative of metaphors and symbols. At the same time, parallel to the decorative and hierarchical functions, in Chinese building tradition; the surface of the wood is painted with these brilliant colors and patterns also protects the material against weathering, corrosion, pestilence, wear and tear, etc. (Jin, 2019, p. 85).

Figure 1.25. The brightly painted gate leading to Sea of Wisdom Temple on the foothills of the Temple of Buddhist Virtue, Summer Palace, Beijing (photo T. Tomaszek, 2018).

Figure 1.26. Typical, reach painting decoration of the interior of the Drum Tower (erected in 1380 during the early Ming Dynasty) in Xi`an, Shaanxi Province (photo T. Tomaszek, 2018).

Figure 1.27. Traditional, bright paint on the Depositary of Buddhist Scripts at Guangren Lama Temple (Tibetan Buddhist Monastery), located in the northwestern corner of the ancient city wall of Xi`an, Shaanxi Province. The temple was built in 1703. It is on the State Council`s list of China`s major religious temples. It is the only temple of the Gelug School of Tibetan Buddhism in Shaanxi Province and it is used to serve as a Palace Hotel for the Dalai Lamas (photo T. Tomaszek, 2018).

Figure 1.28. Traditional, bright paint on the one of the halls of the Guangren Lama Temple (Tibetan Buddhist Monastery), located in the northwestern corner of the ancient city wall of Xi`an, Shaanxi Province (built in 1703); (photo T. Tomaszek, 2018).

Figure 1.29. Traditional reach painting decoration of the exterior of imperial building; The Hall of Imperial Supremacy (Huangji dian), built around 1771, the main hall in the area of Palace of Tranquil Longevity (Ningshou gong), which the Qianlong Emperor (r. 1736-1795) built for spending his years after his retirement. Forbidden City, Beijing (photo T. Tomaszek, 2018).

Figure 1.30. Traditional reach painting decoration of the interior of imperial building; The Hall of Imperial Supremacy (Huangji dian), Forbidden City, Beijing (photo T. Tomaszek, 2018).

Figure 1.31. Decorative painting of the interior of Pavilion in the Imperial Garden, the Forbidden City, Beijing (photo T. Tomaszek, 2018).

Figure 1.32. Reach painting decoration, the Gate of Supreme Harmony, the Forbidden City, Beijing (photo T. Tomaszek, 2018).

Figure 1.33. Reach painting decoration, the Gate of Supreme Harmony, the Forbidden City, Beijing (photo T. Tomaszek, 2018).

Figure 1.34. Decorative painting, The Gate of Tranquil Longevity (Ningshou men), the Forbidden City, Beijing (photo T. Tomaszek, 2018). The Gate, together with the Palace of Tranquil Longevity was built in 1689. During 1772 – 1776 the area was renovated, and the gate was modified according to the design of the Gate of Heavenly Purity (Qianqing men); (photo T. Tomaszek, 2018).

Figure 1.35. Decorative painting in one of the rooms of the Potala Palace, Lhasa, Tibet. The construction of Palace started in 1645. It was the winter palace of the Dalai Lamas from 1649 to 1959, has been a museum since then, and has been a World Heritage Site since 1994 (photo T. Tomaszek, 2018).

Figure 1.36. Decorative painting of the interior of Direpu temple (Dirapuk monastery) located on the foothills of Mt. Kailash, Tibet (photo T. Tomaszek, 2018).

Figure 1.37. Decorative painting of the interior of Zuchupu temple (Zongzhepushi Temple), located on the foothills of Mt. Kailash, Tibet (photo T. Tomaszek, 2018).

According to Liu (Liu, 1984) and Ma (Ma B.J., 2003) the traditional Chinese timber structure construction system can be analyzed with regard to the construction frame design solutions and geographic features. Therefore it can be divided into three types: a through-type frame, a beam-lifted frame and a log-cabin-type frame (Figure 1.38 – 1.64).

1) Through-type frame. This type of frame is mainly used in rural housing and thus there are hardly any references for this type of construction in official building standards. It is constructed of a vertical connection with a separate frame. Columns are usually joined by square crossbeams along the entire length of the house to form a truss, and then by square crossbeams to connect every two trusses to form the frame of the house. The advantage of this type of structure is the use of materials with small cross section that are easy to obtain, as well as the use of multiple square crossbeams along the length of house which can be assembled in advance, increasing the entirety and stability of the structure. Due to its simple practice, direct force transmission and its ever-evolving and adaptable nature, the installation of the walls is very convenient, and saves manpower as well as materials.
2) Beam-lifted frame. This type of frame was formed during the Spring and Autumn periods, then continuously evolved over centuries to eventually become a settled practice. It was widely used in northern China in larger buildings such as palaces and temples. The Beam-lifted frame typically consists of a frame layer, the Dou-gong bracket layer, and also a roof layer. The structure is constructed by placing a beam head on top of a column and then on top of it, using a shorter column to hold a shorter beam and another beam head and another column, and so on. Finally, the short column on the short beam carries the weight of the purlin. Buildings constructed of this frame type differ in the size, the amount of the material and the combination of frames depending on the social rank of the user. These construction rules, as well as the others, have been

strictly defined in regulations such as building standards in the Song Dynasty and construction practices in the Qing Dynasty. A characteristic feature of the beam-lifted frame is the large distance between the columns along the entire length of the building as well as enclosed a larger interior space. Additionally, constructions of this type are characterized by special aesthetic values.

3) Log-cabin type. An ancient type of structure that dates back to the primitive society. According to historians, this type was used in building an outer coffin in Shang Dynasty tombs from more than 3,000 years ago and in the caved patterns on Han Dynasty relics found in Yunnan province in southwestern China. In buildings of this type, it is possible to regulate the room temperature and thus adapt to the changing climate in mountain areas. At the same time, the structure is largely earthquake resistant. The log-cabin type of building, requires only basic materials and minimum manpower, however, it is quite versatile and mobile. The disadvantage however, is that a large amount of wood is needed to build this type of house, and the size and location of doors and windows are greatly limited. This type of construction is most common in Inner Mongolia, the dense forest areas of northeast China, and mountainous areas in Sichuan Province and Yunnan Province in southwestern China. In the northeast part of China, it is referred as "Mukeden", meaning to build houses from piled up carved logs (often cut into semi-cylinders). This type of construction is not as widespread in China as the other two types.

Figure 1.38. Traditional wooden houses (through-type frame construction) in Longsheng Dazhai Red Yao Minority Village, Jinkeng Rice Terraced Fields (part of Longji Rice Terraces), Longsheng County, Guangxi (formerly a province, an autonomous region from 1958); (photo T. Tomaszek, 2018).

Figure 1.39. Traditional wooden bridge (through-type frame construction) in Longsheng Dazhai Red Yao Minority Village, Jinkeng Rice Terraced Fields (part of Longji Rice Terraces), Longsheng County, Guangxi (formerly a province, an autonomous region from 1958); (photo T. Tomaszek, 2018).

Figure 1.40. Traditional wooden bridge (through-type frame construction) in Longsheng Dazhai Red Yao Minority Village, Jinkeng Rice Terraced Fields (part of Longji Rice Terraces), Longsheng County, Guangxi (formerly a province, an autonomous region from 1958); (photo T. Tomaszek, 2018).

Figure 1.41. Traditional wooden houses (through-type frame construction) in Longsheng Dazhai Red Yao Minority Village, Jinkeng Rice Terraced Fields (part of Longji Rice Terraces), Longsheng County, Guangxi (formerly a province, an autonomous region from 1958); (photo T. Tomaszek, 2018).

Figure 1.42. Traditional wooden houses (through-type frame construction) in Longsheng Dazhai Red Yao Minority Village, Jinkeng Rice Terraced Fields (part of Longji Rice Terraces), Longsheng County, Guangxi (formerly a province, an autonomous region from 1958); (photo T. Tomaszek, 2018).

Figure 1.43. Lexu Hall, also named Hall of All Families (through-type frame construction), built during the Ming Dynasty in 1403, Hongcun village, Yi County, Huizhou region, southern Anhui Province (photo T. Tomaszek, 2018).

Figure 1.44. Lexu Hall, also named the Hall of All Families (through-type frame construction), built during the Ming Dynasty in 1403, Hongcun village, Yi County, Huizhou region, southern Anhui Province (photo T. Tomaszek, 2018).

Figure 1.45. Nan Hu Shu Yuan (Intellectual Family School), also referred to as South Lake School, (through-type frame construction), built during the Qing Dynasty in 1814, Hongcun village, Yi County, Huizhou region, southern Anhui Province (photo T. Tomaszek, 2018).

Figure 1.46. Typical wooden houses (through-type frame construction) at Zhaoxing Dong Village, Guizhou Province. The village, belonging to the ethnic minority of Dong, was built in 1160A.D. (in the 5th year of the Southern Song Dynasty). It is the most famous of the group of eight Dong Villages, called sometimes by similar name as "Zhaoxing Dong Villages" (photo T. Tomaszek, 2018).

Figure 1.47. Traditional wooden houses (through-type frame construction) at Zhaoxing Dong Village, Guizhou Province. The village, belonging to the ethnic minority of Dong, was built in 1160A.D. (in the 5th year of the Southern Song Dynasty). It is the most famous of the group of eight Dong Villages, called sometimes by similar name as "Zhaoxing Dong Villages" (photo T. Tomaszek, 2018).

Figure 1.48. Traditional wooden building (through-type frame construction) at Zhaoxing Dong Village, Guizhou Province. The village, belonging to the ethnic minority of Dong, was built in 1160A.D. (in the 5th year of the Southern Song Dynasty). It is the most famous of the group of eight Dong Villages, called sometimes by similar name as "Zhaoxing Dong Villages" (photo T. Tomaszek, 2018).

Figure 1.49. The Hall of Preserving Harmony, also called the Hall of Preserved Harmony (Baohe Dian), Forbidden City, Beijing. Like most of the Forbidden City built during the Ming Dynasty (1368-1644) in 1420 by Emperor Yongle (1402-1424). Like several other halls it perished in the 1625 fire and was subsequently rebuilt. It was renovated again in 1765. It is the beam-lifted frame construction (photo T. Tomaszek, 2018).

Figure 1.50. The Hall of Union (Jiaotai dian), originally named the Hall of Central Perfection (Zhongyuan dian), built in the Ming dynasty in 1535. This hall contains the seals of the Qing Dynasty and the clocks that set the time for the imperial city over the years. It was restored in 1665 and again in 1669. It was rebuilt after a fire in 1798 and exists now as it was built then. It is a beam-lifted frame construction (photo T. Tomaszek, 2018).

Figure 1.51. Interior of the Hall of Union (Jiaotai dian), originally named the Hall of Central Perfection (Zhongyuan dian), built in the Ming dynasty in 1535. This hall contains the seals of the Qing Dynasty and the clocks that set the time for the imperial city over the years (photo T. Tomaszek, 2018).

Figure 1.52. The halls of the Jade Buddha Temple in Shanghai. Temple was founded in 1882 to house two jade Buddha statues brought to Shanghai from Myanmar. During the revolution that overthrew the Qing Dynasty, the temple was destroyed, but the Buddhas were saved. In 1928, a new temple was built for them on the same site and was named the Jade Buddha Temple. It is a beam-lifted frame construction (photo T. Tomaszek, 2018).

Figure 1.53. The Grand Hall of the Jade Buddha Temple in Shanghai. Temple was founded in 1882 to house two jade Buddha statues brought to Shanghai from Myanmar. During the revolution that overthrew the Qing Dynasty, the temple was destroyed, but the Buddhas were saved. In 1928, a new temple was built for them on the same site and was named the Jade Buddha Temple. It is the beam-lifted frame construction (photo T. Tomaszek, 2018).

Figure 1.54. Interior of the Grand Hall of the Jade Buddha Temple in Shanghai (photo T. Tomaszek, 2018).

Figure 1.55. Simple log dwelling with log outbuildings in Da Luoshui Village on the Lugu lakeshore, Yunnan Province. This village is a home to Mosuo ethnic minority group famous for its matriarchal society and their wooden log architecture. Lugu lake, an alpine lake at an elevation of 2,685 metres, is located on the border of Yanyuan County, Sichuan Province and Ningland Yi Autonomous County, Yunan Province. The lake`s shores are inhabited by many ethnic minority groups, such as the Mosuo, Norzu, Yi, Pumi and Tibetan. The most numerous of these are the Mosuo people (photo T. Tomaszek, 2019).

Figure 1.56. Simple log dwelling in Da Luoshui Village on the Lugu lakeshore, Yunnan Province (photo T. Tomaszek, 2019).

Figure 1.57. Simple log dwelling in Da Luoshui Village on the Lugu lakeshore, Yunnan Province (photo T. Tomaszek, 2019).

Figure 1.58. Simple log dwelling with log outbuildings in Da Luoshui Village on the Lugu lakeshore, Yunnan Province (photo T. Tomaszek, 2019).

Figure 1.59. Traditional log houses in Da Luoshui Village on the Lugu lakeshore, Yunnan Province (photo T. Tomaszek, 2019).

Figure 1.60. Traditional log houses in Da Luoshui Village on the Lugu lakeshore, Yunnan Province (photo T. Tomaszek, 2019).

Figure 1.61. Traditional log houses in Da Luoshui Village on the Lugu lakeshore, Yunnan Province (photo T. Tomaszek, 2019).

Figure 1.62. Traditional log structures in the remote Tongle village (near Yezhi), Yunnan Province (photo T. Tomaszek, 2019).

Figure 1.63. Traditional log structures in the remote Tongle village (near Yezhi), Yunnan Province (photo T. Tomaszek, 2019).

Figure 1.64. Traditional log structures in the remote Tongle village (near Yezhi), Yunnan Province (photo T. Tomaszek, 2019).

There were extensive field studies in the 1980's carried out on the structure materials of historic wooden buildings in China. The majority of the research took place in 11 provinces, municipalities and autonomous regions in China[3]. Part of these studies was devoted to identify the species of wood traditionally used for these historic structures (Ze-li, et al., 2017, p. 204). According to the field researches and some microstructure identifications, the species of wood for the main components of the historic buildings such as beams, fangs (a square pillar), columns, purlins and rafters vary slightly according to the region of China. And so in southern China they were mainly made of nanmu (*Phoebe zhennan S. Lee*), cypress (*Cupressus funebris Endl.*), China fir (*Cunninghamia lanceolata (Lamb.) Hook.*) and pinus (*Pinus massoniana Lamb.*), whereas in northern China of Chinese pine (*P. tabulaeformis Carr.*), larch (*Larix gmelinii (Rupr.) Kuzen.*) and pinus armandi (*P. armandii Franch.*). The difference in wood used, is also noticeable according to the type of building. In common housing, poplar (*PopulusL.*) and elm (*Ulmus pumila L.*) are used, while in large-scale historic buildings of great importance, there are wood materials such as nanmu from the south involved. It indicates that these buildings were mostly constructed using local materials. The exception being in the buildings of greater importance, where higher standards were applied. Additionally, buildings in Sichuan province in southwest China and Hubei province in the central-south are constructed of nanmu and cypress for there was a large reserve at the time (Ze-li, et al., 2017, p. 204).

3. As a result of these extensive field studies a national mandatory standard called: *Technical code for maintenance and strengthening of ancient timber buildings*, was established in 1993. (*Technical code for maintenance and strengthening of ancient timber buildings,* National Technical Committee 41 on Timber of Standardization Administrator of China, GB 50165-92; China Building Industry Press; Beijing 1993.)

The significant remaining historic wooden buildings in China – some examples and characteristics

ABSTRACT: It is believed, that Chinese wooden architecture is the most important source of inspiration for Asian architecture. Some of the oldest and most magnificent wooden structures in the world have been preserved to this day in China. The most significant among them are temples. These display not only the traditional construction solution that has been in use in China for a thousand years, but also a deep spiritual dimension and philosophical thought that has shaped the history of Asian architecture. In this chapter, they are presented and characterized in detail. Some of the oldest and the most significant remaining wooden structures in China; namely the Foguang, Nanchan, Dule Temples and the Yingxian Wooden Pagoda of Fogong Temple. The detailed description of these buildings is enriched with photographs showing their outstanding beauty, highly advanced technological solutions and reach paint decoration. This section is supplemented with information of past conservation works applied during tangled times of Chinese history.

The richness of traditional architectural forms and construction achievements of Chinese wooden architecture is manifested in the variety of types of local construction and in almost endless variations of individual building solutions. This wealth is best reflected in the refined historic vernacular buildings scattered throughout the country. An example of this, is the log structures that captivate with their beauty, squeezed between the steep slopes of the endless mountain ranges of the Yunnan province. In the case of these buildings, we can admire specific forms or solutions implemented in a given region for generations, implemented in the preserved structures as the quintessence of accumulated technical knowledge and aesthetic taste acquired over generations. Due to a number of factors, whether economic, historical or political, it is difficult to find preserved buildings of common vernacular architecture from those more distant epochs, when the characteristics of today's building were just taking shape. Those that have survived have already appeared as another, secondary materialization of the living building tradition, and thus inform us only indirectly about the forms and structures that are slowly falling into oblivion. Among the oldest preserved objects of wooden, the buildings of a higher rank and more important function dominate, such as sacred objects, for example temples. Those, that naturally have more sophisticated construction and shapes, can be defined as representing "traditional Chinese architecture" in the strict sense of this meaning.

This is, in a way, a typical situation, because for the subsequent past millennia, average buildings of everyday use were not treated as those whose preservation is for some reason necessary. Luckily, in China a few wooden buildings survived. These structures are very old, dating back to ancient times. They are among the oldest wooden structures in the world, and many people believe that they may even be the oldest. It is thanks to them that we can directly admire the architectural shapes and construction solutions that we do not see any more in modern times. Among the most significant remaining historical wooden structures, it is impossible not to mention four special places, which are the pride of the Chinese people. These buildings are: Foguang Temple, Nanchan Temple, Dule Temple and Yingxian Wooden Pagoda of Fogong Temple.

1 FOGUANG TEMPLE

The Foguang Temple (in Chinese: 佛光寺) is the Buddhist shrine erected in wood and located in one of China's four sacred mountains for Buddhism – Wutai Shan mountain in Shanxi Province. Buddhism has been respected in China since the Sui and Tang Dynasties, and by the middle of the Tang Dynasty this religious complex had reached its peak. During this time, the Wutai Shan area has started to be recognized as a holy place for Buddhism. The accumulation of thousands of years has led it to be surrounded by large and small temples (Xia, 2013, p. 7).

The Foguang Temple, is a tribute to the peak of Chinese Buddhist art and architecture. In its architectural design the temple shows early Indian architectural elements and it is built in a remote place on multiple levels following the slope of the sacred Wutai Mountain, among a forest of pine, fir, poplar and willow trees. The temple is oriented in an east-west position, which is in contrary with the typical layout. Chinese temples are traditionally oriented in a south-north position. It is due to its location, at the foot of the mountains, which are surrounding it from the east, north and south (Steinhardt, 2004, p. 233). This particular location, with mountains behind, is believed to improve its Feng Shui and thus it has great significance (Bramble, 2003, p. 115).

The Foguang Temple was forgotten for centuries, and then tracked down in 1937 by modern China's earliest preservationists and architectural historians Liang Sicheng and Lin Huiyin (founders of the Society for Research in Chinese Architecture) when they set off for Wutai Shan to search for ancient buildings (Steinhardt, 2004, p. 228). They were guided by a Tang Dynasty wall painting from the 10th century that was painted in cave 61 of the Mogao Grottoes in China's northwest, which depicted the location of the Foguang Temple on the Wutai Mountain. This painting indicated that Foguang Temple was an important stop for Buddhist pilgrims in the past centuries, thus as a site it had particular significance (Steinhardt, 2004, p. 237).

According to archival information, the Foguang Temple was founded in the fifth century. This was during the Northern Wei Dynasty (386–534 AD, one of the Northern Dynasties) during the reign of Emperor Xiaowen (471-499 AD), and hence the name means 'Buddhist Halo', most likely it purportedly witnessed halos of lights in the sky. From the years of 785 to 820, it underwent several building phases, and a three level, 32 m tall pavilion was built (Wilkinson, 2000; Chai, 1999, p. 83). The architectural complex of the temple was then destroyed in 845 AD (burned to the ground) during Tang Emperor Wuzong's campaign to eliminate Buddhism and foreign influences (Wilkinson, 2000). Fortunately, this jewel of ancient Chinese civilization was rebuilt and enlarged twelve years later. Today, the temple, apart from some smaller halls (Figure 2.1 – 2.2), consists of two main halls: the northern hall called The Hall of Manjusri (Figure 2.3 – 2.6) and the largest hall, the Great East Hall (Figure 2.7 – 2.11). Inscriptions discovered by Liang Sicheng and Lin Huiyin dated the Great East Hall (the major hall of the temple built on the former site of a three storey pavilion) to 857 AD (Qin, 2004, p. 335), confirm that it was re-built on the original layout during the Tang Dynasty (618–907 AD) (Fairbank, 1994, p. 96). Therefore, the Great East Hall, is the second earliest remaining timber structure in China (Steinhardt, 2004, p. 230) [after the older hall at Nanchan Temple, dated to 782, which was also discovered by Liang Sicheng and Lin Huiyin one year later after their re-discovery of the Foguang Temple (Fairbank, 1994, p. 95)] and one of the most important remaining wooden architectural jewels of ancient Chinese civilization (it is one of two last remaining Tang Dynasty Chinese temples), and thus is considered the "primary national treasure" (Ze-li, et al., 2017, p. 202-203). The temple also contains another historically significant architectural wooden structure – mentioned above the northern hall, called Manjusri Hall, dating from 1137 (Jin dynasty) (Qin, 2004, p. 342). Additionally, on the temple grounds, the second oldest existing pagoda in China (after the Songyue Pagoda) has survived, dating from the 6th century (Steinhardt, 1997, p. 231).

Figure 2.1. The Garan Hall, The Foguang Temple, Wutai Shan mountain, Shanxi Province (photo T. Tomaszek, 2019).

Figure 2.2. Elaborate Dou-gong brackets, The Garan Hall, The Foguang Temple, Wutai Shan mountain, Shanxi Province (photo T. Tomaszek, 2019).

Figure 2.3. The northern hall called The Hall of Manjusri, The Foguang Temple, Wutai Shan mountain, Shanxi Province (photo T. Tomaszek, 2019).

Figure 2.4. The northern hall called The Hall of Manjusri, The Foguang Temple, Wutai Shan mountain, Shanxi Province (photo T. Tomaszek, 2019).

Figure 2.5. Dou-gong brackets of the northern hall called The Hall of Manjusri, The Foguang Temple, Wutai Shan mountain, Shanxi Province (photo T. Tomaszek, 2019).

Figure 2.6. The interior of the northern hall called The Hall of Manjusri, The Foguang Temple, Wutai Shan mountain, Shanxi Province (photo T. Tomaszek, 2019).

Figure 2.7. The Great East Hall, The Foguang Temple, Wutai Shan mountain, Shanxi Province (photo T. Tomaszek, 2019).

Figure 2.8. The Great East Hall, The Foguang Temple, Wutai Shan mountain, Shanxi Province (photo T. Tomaszek, 2019).

Figure 2.9. The Great East Hall, The Foguang Temple, Wutai Shan mountain, Shanxi Province (photo T. Tomaszek, 2019).

Figure 2.10. Dou-gong brackets of the Great East Hall, The Foguang Temple, Wutai Shan mountain, Shanxi Province (photo T. Tomaszek, 2019).

Figure 2.11. The massive roof eaves of the Great East Hall, The Foguang Temple, Wutai Shan mountain, Shanxi Province (photo T. Tomaszek, 2019).

Foguang's sophisticated architectural design became a model for Tang architecture across Asian countries, take Todaiji Temple in Nara, Japan as an example. In 2009, the Foguang Temple, was recognized as China's second oldest surviving wooden structure, and was inscribed onto the UNESCO World Heritage List along with the surrounding monasteries comprising the Buddhist cultural landscape of Mount Wutai.

The Great East Hall, the oldest and most important structure of the Foguang Temple, is located on the far east side of the temple complex, atop of a large stone platform (Figure 2.14). The building of the hall is a massive single storey structure measuring seven bays by four or 34 by 17.7 meters. It is supported in its entirety by inner and outer sets of columns with complicated sets of brackets. Seven different bracket types are represented with the height of half of the column itself (Steinhardt, 2002, p. 116), and which support the roof of the hall. The bracket sets are joined by crescent shaped crossbeams, which built an inner ring above the inner set of columns and an outer ring above the outer columns, covered by a lattice ceiling (Steinhardt, 2004, p. 234). It is a top rank example of classic architecture as well as structural frame and construction technology of the Tang dynasty. The method of its construction is extremely sophisticated, and the Dou-gong components are of the highest order and show the exceptionally advanced design and structural features. The temple has a Y-shaped support system in the beam frame, which was the first of its kind in China. It is worth mentioning that, the Dou-gong components have a cross-sectional size of 210 × 300 cm, ten times larger than the same type of elements coming from the Qing dynasty.

The interior of the hall is adorned with 61 m^2 of wall paintings, inscriptions and other treasures coming also from the time of the Tang Dynasty (Ze-li, et al., 2017, p. 203) and later periods, as well as multiple polychrome sculptures (Steinhardt, 2004, p. 234). The hipped-roof and the extremely complex bracket sets, represent Tang Dynasty stylistics and prove to the Great East Hall's importance as a religious structure (Steinhardt, 2002,

p. 116). The high rank of the Great East Hall indicates that even in the Tang Dynasty it was an important building, and no other buildings from the period with such a high rank survive (Steinhardt, 2004, p. 239).

The Manjusri Hall, erected on an 83 cm high platform, is approximately the same size as the East Hall. This wooden building has three front doors and one central back door, it is covered by a large single-eave hip gable roof supported inside on four pillars and diagonal beams (Steinhardt, 1997, p. 232). The interior walls are adorned with murals painted in 1429 during the Ming dynasty (Qin, 2004, p. 342).

In the second part of 20th century, multidimensional research was carried out on the Foguang Temple, and especially focusing on its East Palace. The studies conducted concerned the spatial form, structural bearing capacity, anti-seismic reinforcement and artistic features/characteristics (Chai, 1986). As part of extensive efforts to protect the palace, in 1954 surrounding local residents were moved, with some basic repairs and reinforcement works were begun. In 1985, local authorities built dams around the temple to protect it from mountain torrents, as well as stone walls, flashings and gutters introduced in order to reduce exposure to moisture (Zhang, 2010).

Even though some of the works mentioned above were carried out during the second half of 20th century, the Foguang Temple in general had been almost untouched since the Tang Dynasty and no major conservation work or repairs since the 17th century were conducted until the beginning of 21st century. As a result, the building has retained much of the historic fabric and design authenticity, but at the same time, it suffered serious neglect, and weathering of the structure among other serious threats. Thus, extensive structural damage from falling rocks and rotting roofing, beams and pillars were threatening irreparable damage. Therefore, the complex comprehensive conservation of the temple was urgently needed. In response to Foguang's severely compromised integrity and stability, the extensive project was started in 2005. To save this rare example as a remaining structure of classical Chinese ancient wooden architecture. The project was lead by Shanxi Province's Cultural Relics Bureau and the Shanxi Institute of Ancient Architecture Conservation and Research (SIAACR) with the support of the Global Heritage Fund and Tsinghua University. The Global Heritage Fund established an International Technical Expert Committee to review all plans and conservation work to ensure the highest levels of conservation excellence. Therefore, international teams of specialists were employed. The Master Conservation Plan was finally approved by the State Administration of Cultural Heritage (SACH). Over the next four years, the advanced in situ works were completed, as were architectural and site surveys, complete measurements, mapping, documentation, structural analysis and threat assessment, conservation planning, and site conservation works. Then the necessary emergency interventions including structural supports for the corners of the Grand East Hall and flood protection around the compound were implemented. This was followed by detailed conservation work (Figure 2.12 – 2.13). Finally, the work was completed to improve access to the site, improving roads and infrastructure, and to facilitate site maintenance and enhanced security (Foguang Temple 2005-2009 Progress Report, 2009).

2 THE NANCHAN TEMPLE

The Nanchan Temple (in Chinese: 南禪寺) is a Buddhist temple located similarly as the above described Foguang Temple on Wutai Shan, a sacred mountain in the Shanxi Province, China. According to an inscription on a beam, its oldest structure – the main hall, named the Great Buddha Hall (Figure 2.15 – 2.16, 2.18 – 2.19) was built in the third year of Jianzhong Tang (in 782 CE during China's Tang dynasty, 75 years earlier than the East Hall of Foguang Temple) (Steinhardt, 2004, p. 235). According to historical sources, it was spontaneously constructed as a humble incense temple as a joint venture between villagers

Figure 2.12. Remnants of painted decoration after conservation, The Entrance Hall, The Foguang Temple, Wutai Shan mountain, Shanxi Province (photo T. Tomaszek, 2019).

Figure 2.13. Interior of The Entrance Hall after conservation –reconstructed on wooden elements the painted decoration imitating wood grain pattern, The Foguang Temple, Wutai Shan mountain, Shanxi Province (photo T. Tomaszek, 2019).

Figure 2.14. The Hall of Manjusri and other temple structures seen from the level of the Great East Hall (from atop of a large stone platform on which the Great East Hall is located), The Foguang Temple, Wutai Shan mountain, Shanxi Province (photo T. Tomaszek, 2019).

of the two villages outside Taiyuan (Zhao, 2010, p. 4). As a result of its remote location and small scale, it is not recorded in the county chronicles and Buddhist scriptures. Therefore, the temple escaped destruction during the Buddhist purges of 845 (Liu; Wang, undated). Thus together with the Great East Hall of Foguang Temple, it is one of the most significant historic architectural structures in China, and one of only two wooden structures remaining from the Tang dynasty (Steinhardt, 2002, p. 114). The temple was re-discovered in the 1950s, by architectural historians Liang Sicheng and Lin Huiyin, one year after their re-discovery of the Foguang Temple (Fairbank, 1994, p. 95). In 1961, the Great Buddha Hall was recognized as China's oldest existing timber building and an extremely important structure in the understanding of Chinese architectural history, and thus it was announced as a key national cultural relics unit and placed under government protection (Ancient buildings in Wutai County-Nanchan Temple, undated). Apart from its architectural value, The Nanchan Temple also contains an original set of seventeen artistically-important and original statues that are lined up on an inverted U-shaped dais (Chen, 1954, p. 11). All the sculptures date from the Tang Dynasty (Howard, et al., 2006, p. 371). There are also ancillary buildings, such as a mountain gate.

The ancient architecture of the Tang Dynasty in Nanchan Temple also includes a stone pagoda and three stone lions. The stone pagoda has five floors. The first floor is surrounded by Buddhist stories, and there is a small pagoda on each of the four corners. On the second floor, there is a statue of Buddha carved in the center of each side, and there are two upper and lower layers on each side, and two statues of Buddha are carved on each layer. From the third to the fifth floor, there are three Buddha statues carved on each side of each floor. One stone lion is 78 cm high, and the other two are 45 cm high.

Figure 2.15. The main hall, named the Great Buddha Hall, of the Nanchan Temple, Wutai Shan mountain, Shanxi Province (photo T. Tomaszek, 2019).

During its history the temple has been repaired in later dynasties, except for the main hall, which is a relic of the Tang Dynasty (Zhao, 2010, p. 4). The Great Buddha Hall, also went through some renovation work. One of the inscriptions on a beam indicates that some repairs were already carried out in 1086 under the Song Dynasty, when all but four of the original square columns were replaced with round columns (Steinhardt, 2002, p. 114). Before the Anti-Japanese War, the bricks in Nanchan Temple were demolished by the Japanese army to build a bunker. During the Cultural Revolution, the Buddha statues in the Nanchan Temple were damaged to a certain extent. In 1966, just a few years after it was rediscovered and it became to be known to the world, The Nanchan Temple was heavily damaged in an earthquake. Due to this it was needed urgent conservation, which was carried out in the 1970s. With the assistance of the government, allocating a significant amount of money for the conservation works at the temple (Zhao, 2010, p. 4); (Figure 2.23 – 2.25). This allowed historians a chance to study the building piece by piece and get a glimpse of all its architectural solutions (Steinhardt, 1997, p. 68). Around the late 1970s, 63 monks from various temples in Mount Wutai were sent to live in the Nanchan Temple. For this reason, the Buddha statues in the East and West South Halls were placed outside the temple. On November 24, 1999, the Nanchan Temple was broken into and the remaining Buddha statues of the Tang Dynasty were cut open, and treasures inside were taken.

The Nanchan Temple faces south and covers an area of 3,078 square meters, wherein the north-south range of the temple is 60 meters long and the east-west is more than 51 meters wide. It has a quadrangular courtyard, divided into east and west courtyards. There are 6 halls in total. In addition to the main building, there are three halls of the Great Buddha, which are originals from the Tang Dynasty (Figure 2.20 – 2.22). The quadrangle courtyard also includes other Ming and Qing architecture (Collection of Researches on the Protection of Shanxi Cultural Relics, 2011).

The humble main hall (The Great Buddha Hall) is a typical Tang Dynasty wooden structure on a square floor plan (Figure 2.17). Its width and depth are all three rooms (it is 9,9 meters deep and 11,62 meters across the front), and the plane is approximately square. The main hall faces north and south (Xia, 2013). It is built as beam-lifted structure with single eaves and nine ridge tops. The main hall uses the surrounding 12 strong pillars to support the huge hip-gable roof (single eaves and nine ridge tops, single eaves the so-called 'Xieshan' top) which arise above four rafters. The pillars are implanted directly into a brick foundation and in the upper parts consisting of five-puzuo brackets (Chen, 1954, p. 12). Both sides of the roof are decorated with owl tails. From this, we can deduce that the timber structure beam frame in the Mid-Tang Dynasty were already using "cai" (栱高) as the material standard for timber structure. The main hall embodies the simple, steady and generous style of wooden architecture in the Tang Dynasty. It does not contain any interior columns or a ceiling, nor are there any struts supporting the roof in between the columns. Therefore based on these features, historians believe that this is a low-status structure. Its multiple wooden construction techniques have become isolated examples (Chen, 1954, p. 12), which are of great significance for later studies on architecture of the Tang Dynasty. At the same time, the structure contains several unique features of Tang Dynasty halls that cannot be found anywhere else, including its longer central front bay, the use of camel-hump braces, and the presence of a yuetai (Steinhardt, 1997, p. 68).

In the middle and rear part of the temple, there is a "concave" altar with a length of 8,4 meters, a width of 6,3 meters, and a height of 0,7 meters. There are seventeen colored sculptures of the Tang Dynasty on the altar: the Buddha statue of Sakyamuni is in the center, and Manjushri and Samantabhadra are on the left and right, including the heavenly kings, vajra, and boys. Around the Buddhist altar are brick carvings of patterns and lotus petals, further examples of artwork from the Tang Dynasty.

Figure 2.16. The main hall, named the Great Buddha Hall, of the Nanchan Temple, Wutai Shan mountain, Shanxi Province (photo T. Tomaszek, 2019).

Figure 2.17. Axonometric drawing of Nanchan temple (source: Li qianlang, "Axonometric drawing of Nanchan temple," *Temples Assembly*, accessed April 2, 2022, https://learning.hku.hk/ccch9051/group-13/items/show/46).

3 THE DULE TEMPLE

The Dule Temple (in Chinese 獨樂寺), also known as the Great Buddha Temple, literally "Temple of Solitary Joy" (also translated as "Monastery of Solitary Happiness"), is a Buddhist temple located in the Jizhou District of a suburban area of the city Tianjin in China. It is one of the three remaining Liao Dynasty monasteries in mainland China (The three major monasteries in the Liao Dynasty were Dule Temple in Tianjin, Fengguo Temple in Yi County, Liaoning, and Huayan Temple in Datong, Shanxi). According to historic sources, the origins of the temple date back at least to the early Tang Dynasty, and it is believed that it was built in the second year of Zhenguan in the Tang Dynasty (628 AD) (Anonymous, 1990). Their existence in the Tang dynasty confirms the fact, that in 755 An Lushan held a rally at the temple, at the onset of his rebellion against the Tang emperor.

There are three theories about the origin of the name of Dule Temple. The first, because the internal support of the Guanyin statue is a towering pear tree, named after the homophonic "Du Pear"; secondly, Buddhists have a clear heart and few desires, abide by the precepts, and enjoy the joy of saving all living beings, thus name Dule Temple; the third that, this temple was the place where An Lushan raised troops to rebel against the oath of Tang

Figure 2.18. Dou-gong brackets, the main hall, named the Great Buddha Hall, of the Nanchan Temple, Wutai Shan mountain, Shanxi Province (photo T. Tomaszek, 2019).

Figure 2.19. Dou-gong brackets, the main hall, named the Great Buddha Hall, of the Nanchan Temple, Wutai Shan mountain, Shanxi Province (photo T. Tomaszek, 2019).

Figure 2.20. One of the auxiliary halls of the Nanchan Temple, Wutai Shan mountain, Shanxi Province (photo T. Tomaszek, 2019).

Figure 2.21. Dou-gong brackets, one of the auxiliary halls of the Nanchan Temple, Wutai Shan mountain, Shanxi Province (photo T. Tomaszek, 2019).

Figure 2.22. The roof eaves and Dou-gong brackets, one of the auxiliary halls of the Nanchan Temple, Wutai Shan mountain, Shanxi Province (photo T. Tomaszek, 2019).

Figure 2.23. The remnants of painted decoration after conservation, the Entrance Gate, the Nanchan Temple, Wutai Shan mountain, Shanxi Province (photo T. Tomaszek, 2019).

Figure 2.24. The remnants of painted decoration after conservation, the Entrance Gate, the Nanchan Temple, Wutai Shan mountain, Shanxi Province (photo T. Tomaszek, 2019).

Figure 2.25. Wooden columns after conservation (after replacing degraded parts), the Nanchan Temple, Wutai Shan mountain, Shanxi Province (photo T. Tomaszek, 2019).

Dynasty, thus the name of the temple could then be also a reference to An Lushan, who was also known as An Dule (Steinhardt, 1997). The last statement is the most widespread (Yao, 2013).

Unfortunately, no any structure of the temple survived from the time of Tang dynasty. The oldest two existing buildings in the temple are two timber-frame structures, the Shanmen (or Shan Gate – the Front Gate) and the the Guanyin Pavilion (the central hall pavilion that houses a colossal clay statue of the Eleven-Headed manifestation of the Bodhisattva Guanyin (Avalokiteśvara). Both of them were rebuilt in the second year of Shengzong Tonghe Emperor of the Liao Dynasty (984 AD), that is, the first year of Emperor Yongxi of the Song Dynasty. The pavilion was again rebuilt in the early years of the Northern Song Dynasty. They are recognized as important representatives of Liao Dynasty architecture (Figure 2.26 – 2.32). These two buildings were designed and constructed by local architects and craftsmen and both structures repeat some architectural solutions from the earliest buildings, as they were constructed on basis of the Tang architectural technology and carving techniques. Both these buildings therefore, are also among the oldest surviving wooden structures in China (Steinhardt, 1997), thus the temple is considered to hold particular historical as well as architectural significance. The statues in the two buildings also come from the same time as the architecture, but like the architecture, they have undergone repairs and alterations by subsequent generations. The other buildings in the Dule Temple, are all from the Ming Dynasty. The construction, including the palace in the east of the temple, was built in the eighteenth year of Qianlong in the Qing Dynasty (1753).

In the 25th year of reign of the Wanli Emperor in the Ming Dynasty (1595), the King of Hubu Langzhong undertook a large-scale renovation of the Dule Temple. According to this record, the repairs of the Dule Temple in the Ming Dynasty were carried out in the late Wanli period. The "reconstruction" recorded in history books was only limited to the oil paintings and polychrome, and the structure of the temples and pavilions of Dule Temple had not been changed during this renovation. At the end of the Ming Dynasty, the Jizhou District was subjected to slaughter by the Qing army on three occasions. According to legend, the people of the whole city gathered at Dule Temple and Taxia Temple, where they were protected from death. Although the city was massacred, the temple was unharmed and the people inside of it survived unharmed. More than sixty years after this event the temple was repaired by Wang Yubi, and then Wang Hongzuo repaired it again, most likely in the fourteenth year of Chongzhen (1614 AD) (Liang, 1932). In the tenth year of Emperor Qianlong, Emperor Gaozong of the Qing Dynasty visited Dule Temple and wrote a book titled "Dule Temple" (Zuyi, 1992). Later, he wrote "Dule Temple – time to rebuild and complete, and the road is approaching [. . .] To teach the Qing to go, I am not the first to worry about Dule" (Zuyi, 1992). In the Qing Dynasty, Dule Temple was once a forbidden place, and civilians were not allowed to enter.

After the Revolution of 1911, Dule Temple was returned to the people, and for a while it was in religious use. However, in 1917, the sixth year of the Republic of China, the west part of the temple's complex was converted into the Jixian Village Normal School.

The temple was used for educational purposes until 1925 when the Shaanxi army came to Jixian and stationed at the Temple's grounds. It was the beginning of the garrison at the temple. Then in 1928 the Jixian Security Force of the Beiyang Government was stationed at Dule Temple grounds, and its main hall was used as barracks for a unit of soldiers commanded by warlord Sun Dianying. The soldiers looted the temple grounds and the decoration of the buildings was particularly damaged. They were also responsible for looting the nearby Eastern Qing Tombs.

For the next couple of months, the destruction of the ancient buildings by the stationed soldiers was becoming more serious. However, compared with the robbery case of the Eastern Tomb of the Qing Dynasty, Dule Temple suffered less damage. The army did not leave their converted military post until the spring of 1929. Leaving behind bullet holes in the timber frames of the Temple that were inflicted during target practice (Steinhardt, 1997).

Figure 2.26. The Shanmen (the Front Gate), The Dule Temple, municipality of Tianjin (photo T. Tomaszek, 2019).

Figure 2.27. The Shanmen (the Front Gate), The Dule Temple, municipality of Tianjin (photo T. Tomaszek, 2019).

After the success of the Northern Expedition, the Chinese Kuomintang's party headquarters in Jixian County were established. From this moment any religious practice, was considered superstition, and was forbidden. So the party committee advocated the auction of Dule Temple. Local people in Jixian County expressed dissatisfaction and opposition, so this proposal could not be realized (Liang, 1932). In 1931, the whole temple was assigned to Jixian Rural Normal School, including Guanyin Pavilion, Shanmen, and East-West courtyards. Both the east and west courtyards and the rear main hall were converted into school buildings, while the Guanyin Pavilion and the Shanmen remained intact (Steinhardt, 1997). In fact, this more than likely helped save the temple during the complicated times of 20th century Chinese history. However, part of the fence on the south side was surrounded by an earthen wall, the local poor people started to slowly tear down the temple structures for wood or other elements. At that time, the principal of the Jixian Country School took the necessary precautions to protect the buildings, but it was obvious that this would not be enough to protect the temple against further destruction. The gate of Guanyin Pavilion was under the management of the school more than ten years, thus even it avoided the total destruction it was seriously damaged (Liang, 1932).

Figure 2.28. Dou-gong brackets, the Shanmen (the Front Gate), The Dule Temple, municipality of Tianjin (photo T. Tomaszek, 2019).

On May 29, 1931, Japanese scholar Guan Yezhen drove to the Tomb of the Qing Dynasty to investigate Jixian County, he accidentally found an old building on the side of the road, spotted through the car window, he stopped and entered through the side door (Yao, 2013). Guan Yezhen, who came to Dule Temple by chance, knew at a glance that it was a very old building from the Liao Dynasty (Sada, 1932). In the same year Liang Sicheng, the famous Chinese architect, historian, and the author of China's first modern history on Chinese architecture, also planned to visit Jixian County. But because his expedition was just over and the current situation was turbulent, he gave up. In 1932, Liang Sicheng's plan to investigate Dule Temple was finally completed (Sada, 1932). After the investigation, academic papers were compiled and published. This made Dule Temple famous both in China and abroad (Liang, 1932).

Figure 2.29. The Guanyin Pavilion (the Central Hall Pavilion), The Dule Temple, municipality of Tianjin (photo T. Tomaszek, 2019).

Figure 2.30. The Guanyin Pavilion (the Central Hall Pavilion), The Dule Temple, municipality of Tianjin (photo T. Tomaszek, 2019).

Figure 2.31. The roof eaves and Dou-gong brackets, the Guanyin Pavilion (the Central Hall Pavilion), The Dule Temple, municipality of Tianjin (photo T. Tomaszek, 2019).

Figure 2.32. Dou-gong brackets, the Guanyin Pavilion (the Central Hall Pavilion), The Dule Temple, municipality of Tianjin (photo T. Tomaszek, 2019).

Figure 2.33. The interior of the Guanyin Pavilion (the Central Hall Pavilion) – noticeable the supporting wooden elements introduced after the earthquake of 1976, The Dule Temple, municipality of Tianjin (photo T. Tomaszek, 2019).

Architect Liang Sicheng once referred to Dule Temple in his book "A Research on the Gate of Guanyin Pavilion of Dule Temple in Jixian County" as "inherited the legacy of the Tang Dynasty, built in the style of the Song Dynasty, it is an important material to study the transformation of architecture in my country (China). It is also a rare treasure" (Liang, 1932). Studies conducted by Liang Sicheng show, that the land of Jixian County (where Dule Temple stands) in the Tang Dynasty belonged to China. Even though at that time this area was located on the border of the Central Plains, thus it was geographically distant and relatively isolated (Liang, 1932), still culture and architecture of this part of the country was directly influenced by the Central Plains.

After the Five Dynasties period (from 907 to 979), the land started to belong to Yidi (name for the traditional 4 groups of barbarians occupying the land outside the borders of ancient China), therefore, if new culture trends emerged in the Central Plains they were efficiently blocked by political boundaries at the time. As the original Chinese culture was solidly conservative, without new ideas it was quite stable in Jixian County over the next decades. Therefore, Liang Sicheng judged that when the Guanyin Pavilion was rebuilt, if there were any new changes in the Central Plains architecture, Jixian County would not necessarily be affected by it, and more of the characteristics of the Tang Dynasty were preserved. Such as the Guanyin Pavilion, it is the transitional form of the Tang and Song dynasties, and it is the key building to research this important transition (Liang, 1932).

Liang Sicheng emphasized the prominent Tang Dynasty style of the buildings when he discovered Dule Temple, and speculated that Dule Temple was founded in the early years of the Tang Dynasty. In view of the fact, that the first Liao dynasty inscription that mentioned the name of Dule Temple, stated that the project, in the second year of Emperor Tonghe (984) was "reconstruction" (building) and "remodeling" (Guanyin image), so the building and sculpture should be created thus before 984. In addition to the overall architectural style of Guanyin Pavilion, mentioned by Liang Sicheng and the strong Tang style of Guanyin Pavilion's two statues, the Liao Dynasty Architecture Research Group of the School of Architecture of Tianjin University discovered several other pieces of evidence in recent years. The establishment that the Guanyin Pavilion was more clearly extended to the Wu Zetian dynasty in the early Tang Dynasty (the end of the 7th century). When the Guanyin Pavilion was repaired in 1998, the carbon 14 dating (tree-ring correction) was carried out on the transferred wood components used in the Guanyin Pavilion. It allowed the discovery that during an important period, namely the "Anshi Rebellion" and "Huichang annihilation of Buddha", the Guanyin Pavilion was also repaired (Liang, 1932).

There have been no less than five detailed studies on the presumption of the basic modular scale of the Liao Dynasty architecture represented by the Dule Temple Guanyin Pavilion. After studying the Dule Temple, Chen Mingda said: "If you talk about the exquisite technology and the quality of art, they should all be ranked first, which can be said to be the top grade among the existing ancient buildings." (Yao, 2013).

In 1961, the Dule Temple was selected as one of the architectural objects included in the first batch of national key cultural relics protection units. That were announced by the State Council of the People's Republic of China, and then approved by the Tianjin Municipal People's Government, as a historical building with a special protection grade (promulgated on March 4, 1961). In 1966, when the Cultural Revolution broke out, large-scale destruction of cultural relics, especially religious buildings, occurred frequently throughout China. Among those who advocated for the preservation of Dule Temple, a key role was played Liang Sicheng, who ventured to Jixian County and proposed to "install lightning rods and install doors and windows for Guanyin Pavilion. To prevent birds from falling on the head of Guanyin, we must cover the head of Guanyin with barbed wire." This prompted the Ministry of Culture to allocate 9,000 yuan that year. The team completed the construction and installation, so that Dule Temple was properly protected (Yao, 2013). In 1972, a mural was discovered during the renovation of the Guanyin Pavilion, which turned out to be covered by a layer of about 1 cm thick ash coming from the eighteenth year of Qianlong (1753) in the Qing Dynasty when the Dule Temple was overhauled. In July 1976, the Great Tangshan Earthquake caused a wall of the Dule Temple to collapse. Part of the wall skin of the Guanyin Pavilion fell off, but the beam frame did not appear to be out of shape, which is quite rare. Since then, the unique seismic performance of Dule Temple, which has withstood many earthquakes without collapsing after a thousand years, has become a research topic in the field of architecture (Jianwei, et al., 2002). In March 1990, the State Administration of Cultural Heritage approved the establishment of the Dule Temple conservation project and listed it as a key ancient national building protection and restoration project. In 1993, the State Administration of Cultural Heritage listed Dule Temple as a key object in the preliminary list for the declaration of World Cultural Heritage. In 1998, the Guanyin Pavilion went under complex conservation works, during which the Tang Dynasty lotus hook head was also found on the top of the Guanyin Pavilion (Dida, 2003, p. 275); (Figure 2.33 – 2.35).

The famous Shanmen building of the Dule Temple, a front gate located on the temple's south side, is a single-story rectangular building which is 10 meters tall. Covered by three single-eave Wudian roofs (roofs with four slopes and five ridges) that has two Chiwens on both ends of the main ridge that are dated to the Liao dynasty (907-1125). The building has twelve pillars and wide opening of the 6,10 meters between the central pillars, which makes the tunnel for entering and exiting the Dule Temple. In its interior, the Shanmen houses the statues of two colored clay statues of guardian kings Heng and Ha sculpted in the Liao dynasty (Steinhardt, 1997, p. 89) and frescos of the Four Heavenly Kings drawn in the Qing dynasty (1644-1911) (Zi, 2012, p. 58). The statues of the two heavenly kings are standing in the two southern rooms, and whether there were statues in the two northern rooms is yet to be verified. The depth between the center pillar and the front and rear eave pillars is 4,38 meters. As the depth is less than the width of the tip space, the intersection of the vertical ridge and the normal ridge is within the tip space and not above the middle column (Liang, 1932). The Shanmen has a stone base, rather low, about 0,5 meters high. The front and back exits are about 2,20 meters, while the two platform exits are 1,30 meters. Obviously, there is no need for pedestrians to go around the gate or pass under the two side exits. There are three levels of stone steps in the south, which are quite short. The stone steps on the north side have been destroyed, and should be the same as on the south side (Liang, 1932). Since the platform base is low, the bucket arch is majestic, and the eaves are far-reaching, and the ridge-end tail is strongly shaped. Giving a solemn and stable feeling. The interior does not have ceilings, and the bucket arches, beams, purlins and other components are all exposed, so in this way the particular decorative effect of the temple is fully brought into full play. Through the bright room at the back of the Shanmen, the Guanyin Pavilion can be observed in the sight range, with neither obstruction nor large gaps. This kind of spatial relationship treatment was obviously designed deliberately by the builders (Liang, 1932). All the trade-offs in the building are very different from the buildings of Ming and Qing Dynasties, and the appearance is very solemn and stable (Dida, 2003, p. 276).

The most impressive structure in Dule Temple complex, the 23 meters in height Guanyin Pavilion (the central hall pavilion), is a three-story timber rectangular structure with five single-eaves Xieshan roofs (half-hipped, half-gabled roofs). It is the earliest existing wooden structure pavilion in China. It is believed to have been built to protect the giant clay statue of the Eleven-Headed Guanyin in the pavilion. Lord Guanyin hangs on the eaves of the "Guanyin Pavilion" plaque, signed "Taibai", which is said to have been written by Li Bai in the Tang Dynasty. This legend has been verified and approved by the historian and cultural relic appraiser Shi Shuqing. The form of the Guanyin Pavilion in the Dule Temple is consistent with the record on the stone stele in the temple, which is saying "upper and lower levels, five east and west, eight north and south". The pavilion has two rows of column pillars around. The architectural structure of the pavilion is separated into three stories by particular type of space arrangement, in which sets *dougong* on the pillars and architraves on the *dougong* in each layer divide building in separated parts (Xiannian, 2001, p. 73). This solution makes it easier for believers to pay tribute to Guanyin from different angles. Additionally, the architrave are placed around the statue and the patio formed in the center is covered with some kind of an octagonal "caisson", which closely integrates the entire interior space and the statue. The second floor has a dark floor, with the appearance of two floors. It is five rooms wide and four deep. For its construction, 24 different dowel and tenon joints were used without using any nails. On the third floor, the main image is covered with algae wells, and the left and right rooms are flat (Dida, 2003, p. 277).

The shape of the pavilion, due to the low base, the pillars of each layer are slightly inclined inward, the lower eaves are built on all sides, and the upper layer is covered with a gentle slope Xieshan-style roof, so that the shape is both strong in the Tang Dynasty and soft in the Song Dynasty. The characteristic of this is an important example of Liao Dynasty architecture (Liang, 1932).

In the center of the Guanyin Pavilion, stands a beautifully shaped eleven-faced Guanyin statue, 16 meters high, passing through the three empty wells in the middle and reaching the roof. There are ten small Buddha statues above it. The Guanyin statue is a treasure of the clay art sculpture of the Liao Dynasty. On both sides of the statue of Guanyin, are smaller statues of his attendants, also made during the Liao dynasty. The colored sculptures and murals of the Liao Dynasty in the hall, are in the same line as the paintings of ladies in the Tang Dynasty. This is a typical example of ancient sculptors applying the images of the people living at that time to the Buddha (Liang, 1932). The clay statue of Guanyin is dated at Liao dynasty (907-1125), but the artistic style is similar to that in the flourishing period of the Tang dynasty (618-907) (Chinese Academy of Architecture, 1982, p. 90). It is the biggest statue of its kind in existence in China and at the same time the largest surviving clay statue in China.

The architectural treasure of the Guanyin Pavilion is built of more than one thousand individual pieces (Steinhardt, 1997, p. 56), which shows a peculiar technique of constructing wooden buildings in ancient China. The thousands of beams, columns and architraves in the pavilion are arranged in an ordered way with high technique, which shows the excellent wooden architecture technology and achievements in the Liao dynasty (Zi, 2012, p. 59). The Pavilion has experienced many earthquakes, but it still stands tall, which is a miracle in the history of architecture.

Liang Sicheng made extensive arguments and specific suggestions on the future protection of Dule Temple in his book titled "A Research on the Gate of Guanyin Pavilion of Dule Temple in Jixian County". He believed that the protection of Dule Temple should first arouse public attention to the importance of Dule Temple's cultural relics. In order for the public to understand the cultural value of Dule Temple, and the essential significance of the pavilion in the history of Chinese culture and Chinese architecture. As for the technical solutions, Sichend recommended that the roof tiles should be refurbished in time to avoid water intrusion and wood decay. Additionally, attention should be paid to fire and lightning protection, and lightning rods should be installed on the roof ridge. Also the defective doors and windows should be constantly repaired, and other building components should be protected as they are, if they are not fully certain. At the same time, Liang Sicheng called for the government to legislate all the necessary protection treatments, and underlined that any large-scale destruction by the army and theft of bricks by tourists must be equally prohibited (Liang, 1932).

After 2000, Dule Temple was damaged by myotis bats, resulting in the continuous occurrence of sawdust and mud in the Guanyin Pavilion in Dule Temple. The excretion of bats also caused large areas of wetness and cracks in the armpits of the Guanyin statue. The murals of the Yuan Dynasty appeared to be peeling like nail holes. As Dule Temple is a wooden structure, the managers of the site conduct fire drills several times every year (Millennium ancient temple…, undated).

Figure 2.34. Ongoing conservation works in the interior of the Guanyin Pavilion (the Central Hall Pavilion), The Dule Temple, municipality of Tianjin (photo T. Tomaszek, 2019).

Figure 2.35. Some of the historical elements from the Temple complex exhibited in one of the auxiliary halls, The Dule Temple, municipality of Tianjin (photo T. Tomaszek, 2019).

4 THE YINGXIAN WOODEN PAGODA

The Yingxian Wooden Pagoda (also known as the Yingxian Wooden Pagoda of Fogong Temple) was built in 1056 AD during the rule of the Liao Dynasty (Figure 2.36). According to researchers, it is the largest and oldest pagoda (and the same high-rise wooden building) in the world, with 67.31 meters in height, 30.27 meters in diameter, and weighing approximately 7400 tons. It is estimated that about 3700 m^3 of wood was used for its construction. It is a multi-story pavilion type of pagoda with an octagonal cross section and nine floors disguised as five. The building has as many as 54 types of Dou-gong brackets with various functions, shapes and sizes. As a result, in China it is often referred to as a kind of "real museum" of Dou-gong brackets. The main body of the building is heavily tilted, as well as the severe twist of column frame of the second and third floors is noticeable. These deformations are most likely the result of many earthquakes in the last hundreds of years, as well as inadequate historical repairs. Measurements carried out in 2010 showed that the total vertical inclination of the pagoda is over 1.25%, with the highest vertical deformation occurring especially in the area of the second story (Ze-li, et al., 2017, p. 201-202).

As one of the most important historic wooden structures in China, the Yingxian Wooden Pagoda has undergone extensive research since the first half of the 20th century, just to mention detailed measurements carried out in the 1930s by Liang et al. The first comprehensive book on the Yingxian Wooden Pagoda was published in 1966. During the 1970's and 80's, a 10-year renovation of the pagoda was carried out by such architectural experts as Yang Tingbao (Chen, 1966, p. 130). It was then that the issue of its partial tilt was recognized and the basic rules and solutions for repairing and strengthening the pagoda were established. Then, in the 1990's the Committee of Yingxian Wooden Pagoda Restoration and Preservation Construction Management was founded. In 2008, monitoring

Figure 2.36. The Yingxian Wooden Pagoda (The Sakyamuni Wooden Pagoda of Fogong Temple), the oldest entirely wooden pagoda in China, Ying County, Shanxi Province (photo T. Tomaszek, 2019).

of the structural stability of the pagoda began and continues to this day. At the same time, since the 1990s, many scientists and their teams have been studying the structural state, damage dispersion, analysis of seismic reactions and material deformation under external forces (Li; Qin, 2005, p. 207). Proper finite element (FE) models were established on the basis of Dou-gong bracket joints and the entire structural system, as well as the load-bearing quality analyses were carried out under lateral load (Chen, et al., 2013, p. 153). Based on the above data, and on the basis of computer-aided design (CAD) drawings and three-dimensional (3D) models, a restoration model for of the pagoda was created (Yu, 2014, p. 47). In recent years, such conservation approaches as a general repair of the framework, full support of the pagoda and raising the support of the upper section have been considered. However, due to the importance, structural complexity and uniqueness of the pagoda, at present only the strengthening and repair of tilted parts and damaged elements on the second and third floors are agreed (Hou, 2014, p. 39).

The significant examples of historic wooden buildings in China described above, are also examples of objects of international interest. Thus, world-renowned specialists were involved in the conservation projects carried out in recent years on these most important buildings. This guaranteed an up-to-date international discourse on the assumed conservation methodologies and the participation in discussions of professionals with different conservation experiences and knowledge. Most importantly, however, the undertaken conservation works, although of course mainly carried out in the context of Chinese culture, were consulted simultaneously in the aspect of current, international conservation concepts and theories. As a result, one can observe a number of conservation activities that fit in with similar conservation works in other regions of the world, and as a consequence, the discourse on maintaining the authenticity of buildings in question goes beyond the aspect of Chinese culture into international perspective.

So is the preservation of the heritage of wooden architecture in China typical also in the case of other objects, those less internationalized and typical vernacular architecture? Can the discourse on authenticity, raised within of the broadly understood tradition and philosophies of Western culture, be transferred in an unmodified way into the context of Chinese culture? Has the very process of recognizing the cultural heritage and its protection followed the same paths, and what follows from this, does the current understanding of the conservation and authenticity of the heritage of wooden architecture have common points in these two various cultural contexts? These are questions require reflection in order to face the concept of authenticity and the issue of its recognition in the perspective of Chinese culture. To do this, it is necessary to examine the very process of the shaping of heritage consciousness and the need for its protection as well as history of development of protection policies in China, together with the possibilities and sense of implementation in the case of Chinese culture of the concept of authenticity that originated and developed in the west.

The birth of the idea of cultural heritage and the history of development of architectural heritage protection policy in China – overview

ABSTRACT: According to archival sources, the recognition of historic buildings in China as an important part of national identity. Thus conscious attempts in preservation date back to ancient times. The Forbidden City is an excellent example, the most important royal complex in the whole of China. It is here that one can observe consistent conservation efforts aimed at preserving the cultural and spiritual heritage. However, it took years after some of the laws regulating the preservation of cultural relicts in China were officially introduced. This chapter presents a brief history of the development of the idea of cultural heritage protection in China and, as a consequence, the history of architectural heritage protection policy in China. The basis of it is further discussed and a new perspective of the understanding of heritage and the public's enthusiasm to study and protect their traditional national heritage in modern day China.

The tradition of recognizing historical buildings in China as an important part of national identity, and a consciousness to preserving and reusing these structures goes back to ancient times. According to Zhu, historic Chinese documents referring to research into cultural relics (and thus important part of it – architectural heritage called in China Architectural Relics) can be dated back to the 17th and 18th centuries (Zhu, 2012, p. 13). As identified by Jin, an excellent example of this is the Forbidden City, the most important royal complex in the whole of China. It was built in 1406 during the Ming Dynasty, and then preserved and reused as their palace by the following 24 emperors (Jin, 2019, p. 86). Numerous craftsmen were required to maintain the buildings, they developed unique restoration techniques, which are the origins of conservation practices in China today.

The first cultural relics conservation law was promulgated in the feudal period. A special importance was assigned to historic architecture. In 1906, during the later Qing dynasty government, document called: *Promotional Measures on Conservation of Antiquities* was published. It was stated that the regional governments are required to investigate all local heritage sites and then report their findings to the Ministry (Liu, 2016, p. 42). A few years later, in 1916, another document *Provisional Regulation on the Conservation of Antiquities* was proclaimed. It included definitions of five recognized categories of antiquities, among them architectural relics such as the tombs of royals and their predecessors, as well as other places of interest (Liu, 2016, p. 42). However, as noted by Jin, it was difficult to talk at that time about the conscious consideration of the protection of historical architecture with respect to heritage value (Jin, 2019, p. 86).

The first official manuscript titled *Yingzao Fashi,* dedicated to the technical aspects of Chinese architecture and craftsmanship, was written by Li Jie in Song Dynasty in the 12th century. It was re-published at the beginning of the 20th century and used as important manual on the theoretical research and practice of protection activities. In that text the special heritage value of ancient architecture (Jin, 2019, p. 86) was underlined. The first legislation issued by a Central Government in China (in 1930) was *The Law on the Preservation of Antiquities.* It brought new regulations on preservation, ownership, excavation, circulation and the export of antiquities (Huang, 2012, p. 23).

In the years that followed, during the Chinese Cultural Revolution, due to political and ideological circumstances, many heritage objects were destroyed. This situation was finally brought to the end with signing the *Notice on Strengthening the Protection of Cultural Relics* in 1974 by the State Council. Here, in that document, the preservation concept aiming on "authenticity" was emphasized for the first time, that was put forward contrary to the idea of reconstruction of the damaged cultural relics (Jin, 2019, p. 88).

The next important step in heritage protection legislation in China, came in 1982 with introduction of the concept of *Historical and Cultural Areas* by the State Council. There it was proposed to include the protection of areas which are not famous, but possessing historic and cultural significance. From that time, the idea of protection not single/individual monuments but also its surroundings, in the other words integrated protection of architectural heritage, was introduced into conservational practice (Jin, 2019, p. 89). Also in the same year (1982), the Standing Committee of the 5th National People's Congress promulgated the *Law of the People's Republic of China on the Protection of Cultural Relics.* From that time, it became to serve as a basic national law, which rose to the level of the National Constitution. This law has been continuously revised in 1991, 2002, 2007, 2013, 2015 and 2017 respectively.

The situation with built heritage changed drastically with the turn of the 20th and 21st centuries. The Chinese government started overwhelming projects of new constructions around the country, including advanced projects to update living facilities. As a result, a large number of unlisted historic buildings and districts were totally destroyed. It soon lead to a new perspective of the understanding of heritage and, as a consequence, it particularly influenced the public's enthusiasm to study and protect their traditional national heritage (Jin, 2019, p. 90).

Fast and steady economic development of China has also brought a lot of significant changes in social, cultural and economic life. As is indicated by Zhu, all of these changes were possible due to the new, open-door policy and rapid modernization, as well as a progressive globalization process. Never on such a large scale, has such massive urban construction and expansion of industrial capacity led to such an increased social mobility. It has provided a new possibility for Chinese people to easily access and appreciate the country's rich heritage (Zhu, 2017, p. 188). As a result of this process, parallel with cultural and economic transformations, the new challenges for protecting and conserving the natural and cultural heritage of the country have appeared.

Globalization, new trends in international issues in the protection of cultural heritage, followed by the development of heritage conservation in China, increased communication and cooperation with international organizations, has all led to China adopting new concepts regarding the protection of its heritage.

However, it soon became obvious that the direct application of international conservation practices in the context of Chinese heritage, would have brought numerous problems. It was realized by domestic heritage protection agencies that these international principles are not applicable to the traditional buildings within China due to their unique nature. So, the concerns over specific conservation principles were the reason for defending and promoting traditional Chinese conservation methods. All of this led to slow development of specific Chinese principles with the view of better understanding and promotion of domestic cultural heritage (Jin, 2019, p. 91). In meantime, it has also gradually introduced the international term "authenticity" in Chinese heritage protection theory. In following years it was slowly recognized, translated and finally "accepted", particularly after China joined the Convention Concerning the Protection of the World Cultural and Natural Heritage in 1985.

The concept of authenticity, which originated in the western world and was derived from Greek and Latin, was already one of the most globally discussed issues regarding the understanding of cultural heritage, its protection, conservation and restoration. While first attempt of authenticity's definition was presented in The Venice Charter in 1965. It was mainly focused on the historical and physical context of a site or a building as the main markers of this value (ICOMOS, 1965), The Nara Document on Authenticity, which was announced thirty years later, pointed out the need of a much broader definition and importance of understanding it as a culturally specific concept (ICOMOS, 1994). In the other words, it introduced a broader definition of authenticity that is sensitive to a cultural context. This was contrary to the previous concept which emphasized above all: originality (Jokilehto; King, 2001, p. 33-35).

Therefore, it was postulated at the time the respect for the cultural and social values of diversified societies when examining the value of authenticity. Also at the same time, the significance of local voices were highlighted in the discourse regarding authenticity. The proposed definition of the term left room for interpretation in different cultural contexts, and according to diverse cultural, political, and economic conditions. Thus from that time, the importance of direct relation between heritage conservation and cultural diversity is underlined, as well as the need of authenticity's interpretation in relation to the exchanges between people from different parts of the world.

The term ‘authenticity’ and attempts of its interpretation and notions from a Chinese perspective

ABSTRACT: When the term of “authenticity” in protection of cultural heritage was introduced to China its interpretation generated a lot uncertainties and tensions. It was mostly because its “international” understanding was ambiguous in a Chinese context. Additionally, Chinese scholars have since the very beginning translated “authenticity” in different ways. Some underline the significance of a building’s original state, as they read authenticity as “original” and “real”. Others, in contrary, criticize above translation, as it limits authenticity only to the original state of architectural heritage and disregards any historic changes of a structure. All resulted in discussions taking on going until today. This chapter summarizes the variety of translations of “authenticity” from a Chinese perspective, at the same time showing the uniqueness of Asian, and thus Chinese, understanding of “original” and “copy” in the case of cultural heritage, as well as the issues resulting from the Chinese approach to authenticity in preservation and interpretation of architectural heritage.

The introduction of the term “authenticity” in China and the interpretation of it following its “international” understanding has generated a lot uncertainties and tensions. It resulted in increased conservation discourse (Zhu, 2017, p. 189). In parallel, it was the reason for many doubts regarding the methods of how the Chinese government implements heritage conservation under its own interpretation of Chinese, as well international heritage guidelines. Furthermore, it also brought many uncertainties on how stakeholders or local people identify appropriate practices for themselves with respect to the concept of authenticity regarded as a key criterion for heritage preservation that was adopted in Chinese cultural context.

According to Zhu, Chinese scholars since the very beginning translate “authenticity” in different ways. These different understandings are used today parallel in academic contexts and government policies (Zhu, 2017, p. 189). One of them, which was proposed by such authors as Xu Songlin (Xu, 2005, p. 20-23), Zhang Jie (Zhang, 2007, p. 80) or Zhang Chengyu (Zhang, 2010, p. 55), use for translation of “authenticity” the Chinese word “yuanzhenxing” (原真性), where “yuan” (原) means “original”, while “zhen” (真) “real” and “trustworthy.” Similarly, this term underlines the significance of a building’s original state. Another translation, mentioned by such authors as Chang Qing (Chang, 2009, p. 118) and Wang Jinghui (Wang, 2009, p. 87) translate “authenticity” with the word “zhenshixing (真實性)”, where “zhenshi” (真實) emphasizes only “real”, “true,” and “verifiable” as the essence of its meaning. The same this understanding points erroneously to translating “authenticity” to “yuanzhenxing” (原真性), as it limits authenticity only to the original state of architectural heritage and disregards any historic changes of a structure (Zhang, 2010, p. 57-58). As a consequence of these different translations of the term, comes inconsistent “Chinese understanding” of the notion of authenticity, that is, whether or not the original state is essential for the evaluation of authenticity.

The word authenticity is also translated in Chinese language by the word 真情性 (Zhēnqíng xìng – the nature of truthfulness) mentioned more in a World Heritage context. However, as noted by Lee (Lee, 2018, p. 58) who was analyzing the meaning of authenticity in conservation practice in a Korean context, it can be also translated as (原形, Yuánxíng – original shape, true shape). According to Lee, this translation contains a closer meaning of authenticity in conservation practice of Asian countries, where particular meaning is connected to original "form". These various meanings of authenticity [真情性 (Zhēnqíng xìng – the nature of truthfulness) and 原形, (Yuánxíng – original shape)] play similar roles in assessing values of heritage (Lee, 2018, p. 58). It creates a conceptual dichotomy between these two above words. Even though both of them contain the meaning of authenticity, each of them has been understood to have different meaning connotations and therefore different applications in practice. Since the semantic scope of both words have been understood as "the ability of a property to convey its significance over time" (Stovel, 2007, p. 21) both have been used to express identified values, and both words form somehow a theoretical tool in the discourse on conservation.

Referring again to Lee, the words Zhēn (真), which is a part of the word *Zhēnqíng xìng*, and Yuán (原) of the word *Yuánxíng* are the closest Chinese characters that reflect the English concept of authenticity: the former means "true, real, sincere and genuine" and the latter "original". Therefore, both translations, *Zhēnqíng xìng* and *Yuánxíng* may be regarded as the closest terms to the notion of authenticity. However, the key understanding of Zhēn and Yuán, lies more within the intangible expressions rather than within material substance, thus through this conceptualization of both words Zhēn and Yuán the intangible aspect gets more important meaning within the authenticity of heritage (Lee, 2018, p. 58).

Connotations of the above words take on a different shape when we look at the original as a contrary to copy, thus when evaluating whether an object is the original or not. In this case, the meaning of Zhēn (真) and the meaning of Yuán (原) are not really the same. In Chinese understanding Zhēn (真) and as well Yuán (原) also are used to express a copy, however, translation in English doesn't allow to understand it to be genuine or original. To get idea of this issue, we need to remember that in tradition of Asian countries, also particularly to China, the process of making a copy or imitating an original has been a well-established genre of genuine painting treated as a creative activity (Lee, 2018, p. 58). There are three other expressions for copying, and thus three different characters showing three different ways of copying. There are as follow: Lín (臨), *Mó* (摹) and *Fǎng* (仿). And here lays the basic distinction, as both Lín and *Mó* are more a way of practice than actually a creative activity, more like the process of understanding a sublime spirit by repeatedly copying a masterpiece. It stays to contrary to the ancient concept of *Fǎng*, which is more to feel and understand the sublime spirit of an original painting and to follow its spiritual and artistic intent in a creative way. So then, how it is conceptualized in Chinese tradition, the painting coming from the *Fǎng* process is no longer a copy or forgery but it is an original. Therefore, it must be stressed, that in this case the actual creation of the painting followed an authentic process, which particularly includes essential intangible values. Significant criteria of becoming Zhēn (真) and Yuán (原) lies simply not in material elements such as the form, fabric or style but instead mostly in immaterial ones such as intention, creativity, and spirit. What interesting, the majority of paintings coming from the *Fǎng* process does not really possess all that much similarity in style and form to the original (Lee, 2018, p. 59).

Therefore, the meaning of the word *Yuánxíng,* which is derivative from the word Yuán (原) understood as original, should not be narrowed to simply the form and design, but its notion needs to be widened to the various qualifying factors, such as materials and substance, use and function, traditions and techniques, but also to location and setting, and spirit and feeling.

The protection of architectural heritage and the meaning and application of the concept of authenticity in China and wider context of Asian culture

ABSTRACT: According to contemporary conservation thought, the preservation of built heritage, thus wooden architecture, should primarily be aimed at adequately maintaining its value and significance, based primarily on its authenticity. Due to cultural specifics, the authenticity of a wooden building in Asian tradition is residing in and beyond materiality, as it is holistically enriched by traditional knowledge systems and rituals. Therefore, it is often underlined, that in the Asian context the protection of material and spiritual values are not necessarily in conflict and the work of heritage practitioners in many instances reconcile them in conservation. This chapter discusses the application of the concept of authenticity in China and wider context of Asian culture, taking into consideration that the preservation of the material remains of a historic wooden building is not necessarily paramount and there is a centuries old tradition of continuous renewal and rebuilding, where the process of redoing itself embodies essential spiritual significance. This particular conservation approach is further discussed from the perspective of an alternate cosmological view and a different concept of time in Asian culture. Resulting in a regenerative view of the relationships between the past and the present, which are circular and continuous.

After the Nara Document on Authenticity was announced in 1994, it become universally accepted that the notion of authenticity has different connotations, and varies in every cultural area or even country. The consequence of these different understandings of authenticity results in different approaches to the significance and values of building heritage as a historic object, and thus to the conservation of this historic object (Galla, 1994, p. 316-317). Later, when in 2001, UNESCO drafted the Universal Declaration on Cultural Diversity (UNESCO, 2001) cultural diversity was recognized to be as equally important as biodiversity, thus authenticity was challenged in a new context. It could therefore, underline the diversity of the issue of authenticity, but also the issue of protection and conservation of heritage in general. This comes as a means of protection and conservation to go through the same progressive changes as culture, following the influences and styles of each era (De Vos, 2018, p. 166). Years later, namely in 2014, The Nara + 20 Document faced anew the notion of authenticity, defines it as: "Authenticity: A culturally contingent quality associated with a heritage place, practice, or object that conveys cultural value; is recognized as a meaningful expression of an evolving cultural tradition; and/or evokes among individuals the social and emotional resonance of group identity" (Nara+20: On heritage practices…, 2014).

Chinese historic wooden architecture in its physical form represents tangible heritage, but its intangible value represented in materialization of the idea goes far beyond the spectrum of authenticity residing in fabric. Issues on preservation of wooden heritage do not only concern the presence, or the absence, of efforts to conserve but also how to conserve with the primacy of authenticity (Zerrudo, 2018, p. 106). It is due to the search for adequately maintaining the value and significance, which are primarily based on its authenticity. Therefore, the authenticity of wooden building in Asian tradition is residing in and beyond materiality, as it is holistically enriched by traditional knowledge systems and rituals (Fadriquela, 2013, p. 43). Consequently, the notions of authenticity then range from the most tangible manifestation to the most intangible expression included in physical form or eternality of place (Zerrudo, 2018, p. 111). Thus, intangible heritage could be considered as thinking and doing, both in constant reliance. This notion of authenticity, expresses more adequately the tradition of preservation of architecture in an Asian context, and thus can serve to capture this tradition in the case of China. It brings somehow a holistic combination of tangible and intangible attributes to authenticity, lightening a bit the obsession to be, or as close as possible to, authentic. Authenticity thus can be understood as a nature of things, which is none other than the state of truthfulness.

In order to make heritage meaningful to all, namely the owners of the heritage as well as tourists, any action of conservation requires a critical balance between authenticity and sustainability. Therefore, the conservation process must include a series of consensus and compromise between experts and communities, between the ideal standard and the real situation of resources and systems. Thus, authenticity needs to be captured as a hybrid or syncretic in its nature and notion (Zerrudo, 2018, p. 115). The transition from the curatorial approach based in the majority on aesthetic or historic values to conservation to a people-centered approach has a significant impact on how conditions of authenticity are understood and met (Araoz, 2013, p. 149). It also needs to be stressed, that protection of heritage and the search for the notion of authenticity in general lies on the collective memory. As Assmann and Czaplicka argue, "Cultural memory works by reconstructing, that is, it always relates its knowledge to an actual and contemporary situation" (Assmann; Czaplicka, 1995, p. 130). That is to say that the notion of authenticity must be seen not as the frozen value and not related to any collective narration, but more as influenced by contemporary meaning and expectations.

It is often underlined, that in the Asian context the protection of material and spiritual values are not necessarily in conflict and the work of heritage practitioners in many instances reconcile them in conservation. As pointed by Wijesuriya, the strong material and technical focus on the conservation of heritage which has dominated Eurocentric doctrinal practice is now very often considered outdated and quite narrow. This approach, developed for over two centuries to sustain its focus on materiality, is very unlikely to be applied successfully to maintain "the full richness of the authenticity", the aim of the Venice Charter (Wijesuriya, 2017, p. 5).

It was effectively challenged in the Nara document, where for the first time authenticity was captured as a relative concept which depends on its socio-economic context (Araoz, 2011, p. 57). Later the Nara Document was integrated into the *Operational Guidelines* of the World Heritage Committee (UNESCO, 2016) and authenticity, being a essential requirement for nominations of heritage sites for inscription on the UNESCO World Heritage List (it is recognized as one of the three mandatory pillars supporting Outstanding Universal Value), is generally understood following these insights. It seems that China stood up to the task of tackling these issues, being one of the very few countries that has integrated the concept of authenticity into official conservation policy and practice.

Recognized as an essential aspect of conservation, authenticity has influenced the field of conservation in China since the 1980s. Approximately around the same time when the Venice Charter (ICOMOS, 1965) was translated and introduced to China, being one of the essential documents regulating to conservation practice (Zhou, 2018, p. 97). The *Principles for the Conservation of Heritage Sites in China*, which were announced in 2000, constituted a serious attempt to combine the Venice Charter with Chinese conservation principles, particularly these firmly rooted in tradition. As an outcome, it presented an outline of material authenticity in historic building conservation (Zhou, 2018, p. 96).

After being introduced in the 1980s, the term authenticity became widely discussed and it took several attempts to translate its meaning to a Chinese context. Yet during the translation of the Venice Charter, it was translated as *yuan zhenxing* (原真性) which means both originality and truthfulness. Thus, it brought the idea of originality being applied to restoration projects. This notion of authenticity was still reflected in the Qufu Declaration from

2005, where attention was placed on the original style, original material, original structure and original craft, it also possessed the scientific, artistic as well as historic values. Finally, all of the attempts to define authenticity were crystallized and issued by China to ICOMOS in 2000 the *Principles for the Conservation of Heritage Sites in China* (Zhou, 2018, p. 97).

After the year 2000, the Chinese conservation practice began to shift its focus from mostly relic-based to now a more heritage-based (treating heritage as a whole system, combining of tangible as well intangible aspects). It had already been marked in the early version of Chinese Principles that "The aim of conservation is to preserve the authenticity of all the elements of the entire heritage site and to retain for the future its historic information and all its values. (. . .) All conservation measures must observe the principle of not altering the historic condition" (ICOMOS China, 2000, p. 4). Then, after updating document, the 2015 version of China Principles states: "The principle of authenticity is the foundation for the conservation of a site's physical remains. All cultural attributes and traditions should be conserved based on the overall understanding of a site's values. The principle of authenticity relates to both tangible and intangible heritage. It is applicable to archaeological sites and ruins as well as traditional architectural sites that are a testament to history, and can also be used as a guide for the conservation of historically and culturally famous cities, towns and villages that retain their original functions, as well as cultural landscapes. Aspects relating to living heritage sites and the diverse range of cultural values that some sites may have are also important components of a site's authenticity and must be preserved in their entirety" (ICOMOS China, 2015, [revised edition], p. 67).

When seeking to maintain authenticity of heritage values through conservation and following management planning, there is high importance to identify and describe where and how the values of the place/site are most clearly expressed. Therefore, the extent to which the material form and substance is essential to the maintenance of cultural and spiritual values or practices must be questioned. Referring to Stovel it can be postulated then, that authenticity is a "measure of degree to which the values of a heritage property may be understood to be truthfully, genuinely and credibly, expressed by the attributes carrying the values" (Stovel, 2004, p. 3). The problem is a result of the concept still being fixed to a Eurocentric perspective of understanding heritage.

In Asian context, thus reflected in Chinese tradition, the preservation of the material remains of a wooden historic building is not necessary paramount and there is a centuries old tradition of continuous renewal and rebuilding, where the process of redoing itself embodies essential spiritual significance. As a consequence, a historic building in which some older elements are obscured by more recent treatments is still understood as authentic, as its materialization represents a revitalizing spiritual process with continuing embodiment of enduring beliefs and practices valued by the community to which it belongs. This approach as a conservation practice has ramifications for an alternate cosmological view and different concept of time. It is common all over Far East Asian countries, take Shinto temples in Japan, where this practice can be observed in its full implementation. It is, in a way, the result of regenerative view of the relationships between the past and the present, which is circular and continuous (Wijesuriya, 2008, p. 75).

This particular concept of time, as circularity endowed in the nature of reality is observed among others in Buddhist tradition. This is also formed in the foundations of Chinese culture. It needs to be remembered that this circularity is permanently inscribed in activities of Buddhist societies, especially religious activities. Therefore, as noted by Wijesuriya, the renewal process or the continuity of religious places is undertaken by them for various reasons such as earning merit (Wijesuriya, 2005). As a consequence, authenticity must be looked at in relation to the renewal, the use of traditions including craftsmanship, techniques and the materials. It should be particularly be connected to the traditional yet changing knowledge systems that sustained these places for centuries, such knowledge is often largely ignored by "professional" conservators.

What seems particularly interesting to mention is the fact, that in Asian tradition some written sources regulating the process of building (and thus somewhat focused on the values currently defined as authenticity) has been known since ancient times. The best example of this could be the *Mayamata,* the ninth century Indian treatise on housing and architecture, which deals with all the facets of gods temple dwellings, from the choice of a site to the iconography of temple walls. In this book a whole chapter is devoted to the topic of renovation works, just to recall the following paragraph: "A temple (may be) ruined, broken down, fallen down, aged as to its materials or decrepit. Those (temples) whose characteristics are perceptible in their principal and secondary elements (are to be renovated) with their own materials. If they are lacking in anything or have some similar type of flaw, the sage wishing to restore them, (must proceed in such a way that) they regain their integrity and that they are pleasantly arranged (anew), this (is to be done) with the dimensions – height and width – which are theirs, with decorations consisting of corner, elongated and other aedicule, without anything being added (to what originally existed) and always in conformity with the advice of the knowledgeable" (Dagens, 1993).

According to Linstrum "religious buildings occupy a special place in the feelings of those who worship there because of what they represent in their lives. They are symbolic as well as personal" (Linstrum, 1991, p. 3). Religious structures are specifically living objects of architecture, particularly wooden Buddhist architecture in China, in case of which circularity and thus continuity it is the backbone of the spiritual meanings contained in them. Citing words of Philippot "as regards to living architecture, it is obvious that lacunae may have to be completed to a much larger extent than justified on purely artistic grounds for archaeological remains" (Philippot, 1976, p. 11). It is also worth to evoke the insights of Silva, who noted: "The living monuments require the forbearance of both devotee and specialist. Either extreme can produce distortion and eccentricity. On the one hand the restoration or rebuilding concept of pilgrims and peasants can move monuments out of their own balance and replace such sites of hallowed dignity with monstrosities of a previous generation. On the other hand, the pure policy maker may be unmindful of the necessary balance and harmony that should exist between science and religion, and thereby make living and religious monuments no more than specimens of scientific interest" (Silva, 1983, p. 14).

It brings the question of authenticity to the rank of an extremely important issue, both in accessing, as well as in interpreting historic architectural structure, particularly wooden ones. Connected deeply to a cosmological view of circular nature of the time, wood as a fabric brings a special meaning to a Buddhist temple and Buddhist object of devotion in Asian context, including China. Thus, a new questions arises – could the various aspects of authenticity be placed in any order, especially in a hierarchical order? How can authenticity be judged, where does it reside, and for whom? Furthermore, is it possible for lost authenticity to be restored? It is important to bear in mind that when talking about a particular instance of authenticity we must identify what we consider as authentic in that individual case. If the chosen target is to maintain authenticity, it must be assured that whatever is meant to be preserved still exists – otherwise it cannot be preserved. Of course, what aspect of authenticity is being sought (authentic fabric? authentic history? authentic appearance? authentic architectural concept? all four? or something else?) all need to be particularly considered. Therefore, it is essential to have a clear intention in the search for authenticity at the outset, otherwise things may get confused (Linstrum, 1991, p. 6). If, as it has been often argued, the authenticity is "a measure of the degree to which the values of a heritage property may be understood to be truthfully, genuinely and credibly expressed by the attributes carrying the values" (Stovel, 2004, p. 3), the problematic still remains a justified clarification of the issue. Thus, for the judged sense to the conservation process, authenticity must be looked through its aspects rather than its notion *per se.* Additionally, it is necessary to remember, that notion of authenticity, even if as a value is applied mostly to the value of a past situation, is constructed in any given time and space context. Therefore, as noted by

David Lowenthal, what counts as authentic has continually shifted in form, space, and time, thus "the criteria of authenticity we choose reflect current views about how yesterday should serve and inform today" (Lowenthal, 1999, p. 8). In the other words, what people value as authentic is an attribute of the here and now rather than the past (Ndoro, 2018, p. 45).

As mentioned previously, it may be postulated that only what still exists can be preserved. How then to justify the sense of restoration, and in the bigger scale the reconstruction, treated as a return to a certain historical situation? It seems that this issue has adopted a specific understanding in Western thought, where even restoration was often justified, the reconstruction at the same time remained questionable and with time acquiring the status of a strongly criticized method. When the first method allowed the recovery of the original appearance, which happened to be somehow lost by accident, decay or any other alterations (Dellheim, 1982, p. 82), the second, often used all over Western countries, was treated with distance. It was even underlined at some point that "nothing but the original structure would bear the stamp of truth" (Prince, 1981, p. 45). To justify any restoration however, the "minimum intervention" was proposed as the principal conservation method. All above ideas remained on the contrary to the cosmological approach to the world as a circular flow of time and the essence of the repetition of phenomena and situations. Thus, these Western philosophies are applicable to ruined cathedrals that are not in use anymore, but useless in every-day conservation, particularly of buildings that are still in use (Linstrum, 1991, p. 6).

While from the point of western conservation theory, reconstructions maybe treated as serving two important functions: experimental research and interpretation (ICOMOS Charter for the Protection and Management of the Archaeological Heritage (ICOMOS, 1990, Article 7), in Asian context shows the nature of time and spiritual value of continuity. That is why it is so difficult to shift to the Chinese tradition suggestions when western theoreticians like John Ruskin or William Morris, who somehow gave the definition of notion of conservation, emphasizing the question of historic time and authenticity in relation to the original object, and the impossibility to reproduce an object that will hold the same significance in another historical, and as well as cultural context (Jokilehto, 1986, p. 8). This unshakable focus on originality at all times it does not allow acceptance of fact that change over time is real, thus this notion of authenticity promotes the fossilization of heritage (Ndoro, 2018, p. 41). All of this leads to the edge of importance of the intangible part of heritage and it brings awareness to the difficulties in transferring the Western understanding of tradition, religiosity and thus the protection of cultural heritage onto the Chinese realms of mentality and spirituality. Therefore, it remains to consider modifying the meaning of the postulated proposals, for example replacing *minimum* interventions to *appropriate* interventions.

Authenticity of historic wooden building – A Chinese perspective and interpretation

ABSTRACT: Traditionally, a wooden building in China was constructed in accordance with ancient recommendations. The same was true in the case of repairing, in which ancient recommendation must be followed in order for the repairs to be acceptably completed. Traditionally, the material aspects of architectural structures in China were regarded as a manifestation of sublime spirits, they could easily be changed or replaced, as long as these changes reflect the intangible attributes of the structure in question. This understanding was changed in the later part of the 20th century, when modern concepts and practices of preserving material remains also came into effect in a Chinese context. By prioritizing material remains and the need of supreme preservation of fabric, China stressed the importance of underlining the national identity through material remains, as historical monuments in their material form visually and tangibly display the pride of a long history and rich culture of the people of the country. This chapter presents the development of Chinese conservation practice with regards to the value of "authenticity", particularly discussing ideas raised by Liang Sicheng, a pioneering and widely recognized figure in the preservation and conservation of Chinese architectural heritage. Thus, it is discussed the concept of "restoring the old as it was" and further the concept of "original state" as well as "original condition" of an architectural relict and its painted decoration, and the use of it in conservation practice in contemporary China.

Traditionally, a wooden building in China was constructed in accordance with ancient recommendations. Thus, it followed an authentic process of erecting it, including particular ceremonial processes and formulated practices. And so, an arrangement of buildings, their style, size and layout, was heavily influenced by Buddhism spirituality and Confucian as well as Taoism philosophy. Additionally, the traditional recommendation of architectural order were provided by an influential set of guidelines coming from the *Chinese Architectural Manual* of wooden structures (Feng, 2012, p. 67). The same was true in the case of repairing wooden buildings, in which ancient recommendation must be followed in order for the repairs to be acceptably completed. Such construction, as well as repair principles, have always affected the way in which the aesthetic value has been appreciated and how it was interpreted in its authenticity in of form of building during the process of repair.

In ancient times, or let's say traditionally, material aspects of architectural structures in China were regarded as a manifestation of sublime spirits. At the same time incorporating useful knowledge, useful teachings and respected principles regarding building, but also as an extremely important factor, they were expressions of human values. Therefore, the material aspects could easily be changed or replaced, as long as these changes reflect the intangible attributes of this structure (Lee, 2018, p. 59). This understanding was changed in the second part of 20th century, when modern concepts and practices of preserving material remains came into effect also in a Chinese context. By prioritizing material remains and the need of supreme preservation of fabric, China stressed the importance of underlining the national identity through material remains, as historical monuments in their material form visually and tangibly displayed the pride of a long history and rich culture of the people of this country. Such a shift in attitude towards material remains, has in consequence, modified the Chinese perspective towards the relationship between material and intangible expressions. The material substance, has been to this point, a kind of a manifestation of a sublime sprit of religious belief, craftsmanship or an artistic and religious process, it is from this moment reflected in a different semantic scope and it takes priority in the valuation of a historical architectural object. Therefore, it is very often used as essential factor in assessing heritage values, posing a threat to the continuation of traditional values and traditional narration of Chinese wooden building (Lee, 2018, p. 59). At the same time, being at the front of values forming the decision process for conservation, the material aspect of wooden buildings in China also heavily influenced the usage and understanding of the term of authenticity from a Chinese perspective. However, thanks to the Asian tradition of circularity of time and nature of continuity, this material aspect of wooden heritage has not overtaken over the other values and Chinese tradition of building and historical preservation continue to develop their own character.

The early international principles guiding the preservation and restoration of ancient buildings, such as those specified in the Venice Charter, were mostly applicable to historic structures made of durable materials (brick or stone structures). This was recognized soon after the Venice Charter was announced, as it was noticed that not all the recommendations are fully justified in the case of wooden structures. In consequence, it was also an important factor for growing discourse about authenticity at the international forum. In China, where majority of heritage buildings are timber structures, the first theoretical voices started to appear also at that time. Among the most influential architects and theoreticians, Liang Sicheng (梁思成, 1901–1972) should be mentioned, he was a pioneering and widely recognized figure in the preservation and conservation of Chinese architectural heritage. He was also one of the first Chinese professionals who followed the principles of the Venice Charter (Zhan, 2018, p. 91). As noted by Zhu, in 1964, the same year as the Venice Charter was formulated, Liang laid the foundations for a conservational discourse in China (Zhu, 2017, p. 189). He pointed out on the proposition, that: "In restoring architecture that has historic and artistic values, in general we must follow the principle of restoring the old as it was [(修舊如舊xiujiu rujiu)]. According to Liang Sicheng, restoration should be based on the principle of "being restored as it was" and "curing disease and prolonging life" rather than seeking a "brand new appearance" (Liang, 1984 (b), p. 334). However, there may be some difficulties in applying this principle to wooden structures, as it should be less difficult than repairing brick and stone structures" (Liang, 1984 (a), p. 56). Liang's idea of "restoring the old as it was" refers to the original form as an essential aspect of architectural authenticity. He also underlined, that the reinforcement or restoration measures that need to be added should be made as barely visible as possible (Liang, 1984 (b), p. 337). In his later writings, he also said: "we have the responsibility to preserve or restore the original state of the architecture of different historical periods" (...) and "we should do our best to keep or restore the forms when these buildings were first built" (Liang, 2001, p. 27). Following this proposition of Liang, as noted by Nan and Zhang, it is almost impossible to make a clear distinction between new and original members when repairing historic timber structures, because the majority of Chinese wooden structures have been reconstructed as wholes (Lin; Ran, 2002, p. 53).

The conservation approach postulated by Liang, is somehow in contrary to the postulates raised by the Venice Charter, especially those in Article 9, where it is identified that the expected aim of the restoration process is "to preserve and reveal the aesthetic and historic value of the monument and is based on respect for the original material and authentic documents" (ICOMOS, 1965, Article 9). In order to preserve the authenticity of a structure, as The Venice Charter is suggesting, it is necessary to avoid any assumption during restoration and any necessary additions must be distinguished from the original by bearing a contemporary stamp. For the last few decades, the concept of authenticity has become a key factor in defining the approach to conservation, as well as influencing international legislation on protection of heritage. However, it should be understood that authenticity may define a valid approach in a given time, context and specific typology of cultural heritage. Therefore, the approach itself may vary depending on many aspects and through historic time. The test of authenticity, conducted on wooden heritage, should not only be used as an assessment of value, but also as a means of developing a changeable approach to conservation regarding variable aspects and criteria. Recently, some Chinese professionals have begun to distance themselves from the principles of the Venice Charter, even to the point of challenging these principles (Ye; Dai, 2002).

In other Asian countries, the voice of importance of maintaining the original form is also evident. Take Korea as an example, here enacted in 1962 and enforced to date, the so-called "Cultural Heritage Protection Law". In Article 3 (Basic Principles of Protection of Cultural Heritage) it is postulated that "The basic principle for the preservation, management and utilization of cultural heritage is to preserve it in its original state". Therefore it can be said, that the essential principle for the restoration of architectural heritage, including wooden buildings, lies in maintaining an original form (Kim, 2018, p. 68). However, it is also postulated, that the valid contributions of all periods to the construction/building must be respected and maintained. Accordingly, the criteria for authenticity in heritage is heavily grounded in the maintenance of the original form, with respect to all the resulting historic changes (Kim, 2016, p. 28). This understanding is directly derived from the assumption that once the original form is destroyed cultural heritage is already lost and irreplaceable, so the values and authenticity of heritage should be preserved given that the original form is guaranteed when the values and authenticity of the heritage are maintained (Kim, 2018, p. 69). Giving that authenticity must be strictly applied in compliance with the principles of 'maintenance of original form', but at the same time respecting the modifications of all historical periods. However, this needs to be adopted to the particular cultural context, as methods of restoration depends on the natural environment, cultural and historical backgrounds, economic power, and restoration policy formed through all such factors in a given country (Kim, 2016, p. 29). In addition, the maintenance of such understood original form faces a limitation resulting from the gap between the ideal original form and that in reality under the practical conditions. These conditions, according to Kim, in general are: (i) replacement of the materials harmful to the human body; (ii) physical limit to the lifetime of the materials used for a cultural heritage; (iii) level of restoration techniques/materials accumulated to date; (iv) cultural heritage utilization plan and perspectives based on humane study; (v) budget and time period for the restoration; and, (vi) different natural environments of each country. Therefore, there are clear difficulties in quantifying authenticity (Kim, 2016, p. 31–31). "Although restoration methods for wooden architectural heritages like this may basically appear to be similar each other, depending on the preservation policy, natural environment, etc. of each country, significant differences may exist, which are closely connected with authenticity of a cultural heritage" (Kim, 2016, p. 36).

In the year 1999, after years of international discussion on authenticity and the problems connected to its understanding proposed by the Venice Charter, and after the Conference on Authenticity in Relation to the World Heritage Convention in Nara held in 1994, the new charter was adopted by ICOMOS, the Principles for the Preservation of Historic Timber Structures. In the document it was clearly stated, for the repair of wooden buildings, and when "it is an appropriate response to the need to replace decayed or damaged members or their parts", replacement timbers can be used. However, this replacement must be done "with respect to relevant historical and aesthetical values". Additionally, what is important from a characteristic of wood point of view, "it should be accepted that new members or parts of members will be distinguishable from existing ones." (ICOMOS, 1999 (c), Paragraph 9 and 10).

According to Zhu, these new "international" principles, when taken from the perspective of a "Chinese framework," are in reality some kind of concession (a combination) to both Liang's assertion of the principle of "restoring the old as it was" and postulations raised by the Venice Charter. The postulations included the Principles for the Preservation of Historic Timber Structures to allow to some extent to follow an old Chinese practice that waives the need to distinguish between new members and original members of historical timber structures (Zhu, 2017, p. 190).

After the concept of authenticity had been adopted by the Chinese government, it was continuously interpreted, implemented, and adjusted to a Chinese context on the basis of its links with Chinese culture, history, and institutions. The necessity to address the complexity of the concept, was the reason for the creation of a set of "official conservation guidelines". Among them two documents stand out as the most important, which are the Law of the People's Republic of China on Protection of Cultural Relics (The Law of the People's Republic..., 2007) and the Principles for the Conservation of Heritage Sites in China (developed by the Chinese National Committee of ICOMOS and with the help of The Getty Conservation Institute (ICOMOS China, 2004) and then approved by the Chinese State Administration of Cultural Heritage (SACH), known generically as the "China Principles" (Lai, et al., 2004). In the latter document "China Principles", the term yuanzhuang (原狀) was implemented by the Chinese government. This referred to the Liang idea of "preserving or restoring the original state", understood mainly as the maintenance of building form (Agnew, et al., 2004, 41). "China Principles" therefore, link the "international" concept of authenticity proposed in the Venice Charter, which is the underlying respect for original fabric as an authentic document of the past, with the key term of "the original state" (yuanzhuang), translated as "historical condition,", the term which first was introduced in Article 21 of the Law of the People's Republic of China on the Protection of Cultural Relics. The term "yuanzhuang", that has been central to discussions on protection of heritage sites in China, is thus understood as the condition of a site through historical time, and as proposed by Zhu, is referring to the following conditions: (1) the condition prior to any conservation intervention; (2) the condition after being subjected to interventions which was judged to be significant (as for example treatments, adaptations, reconstructions) during the course of its history; (3) a ruined state revealing important historical attributes; (4) the reinstated condition after the restoration of elements that were buried, deformed, partially collapsed, braced, or incorrectly placed, where the original components and form of the structure exist; (5) the historical condition of a setting that is of significance to the site (Zhu, 2017, p. 191). Such proposed understanding of "yuanzhuang" is in accordance with the Venice Charter, where it is indicated that "The valid contributions of all periods to the building of a monument must be respected, since unity of style is not the aim of a restoration" (ICOMOS, 1965, Article 11) and opposes the concept of restoration to an earlier state or rejuvenation.

Somehow in response to an ongoing discourse on methodology of heritage conservation in China, the *Symposium of Contemporary Architectural Heritage Scholars* was held in 2005 in Qufu. This meeting resulted in the adoption of *The Declaration of Qufu* —a *"Consensus on the China-specific Conservation Theory and Practices of Historic Buildings"*, which was developed with the intention to serve as a specific principle for architectural heritage conservation particularly in a Chinese cultural context. The Chinese have developed an art of preserving ancient buildings. China's ancient craftsmen have accumulated a wealth of practical knowledge, this document was therefore based on these centuries-old experiences of traditional Chinese heritage protection practices (Jin, 2019, p. 91).

In The *Qufu Declaration* it is underlined, that it is advisable for Chinese heritage practice to benefit from the international advanced experiences and developed concepts in conservation, but such concepts and experiences "have to be analyzed, digested and absorbed according to the actual situation in China" (Luo, et al., 2005). It is also stated that ancient Chinese architecture that used a predominantly wood structure is very unique. It is characterized by distinctive features which are significantly different from Western architecture. It applies not only to materials, technology and construction, but also damage characteristics as well as protection methods (Luo, et al., 2005). Additionally, these vast differences between Chinese and Western principles resulting from geographic environment, culture, artistic and techniques are the reason for the essential differences in architectural heritage preservation concepts and practices. Therefore, it is necessary to set rules such as principles, policies and methods with particular respect for Chinese building tradition, its features and the unique status quo (Jin, 2019, p. 92).

As it is stressed in the Article 11 of the *Qufu Declaration,* the protection of architectural relics not only aims to protect the relics themselves, but also their traditional culture, materials and crafts. Otherwise, there can be no way relic protection at all (Luo, et al., 2005). In actuality, this understanding of architectural relics protection recognizing importance of intangible cultural heritage, to which are actually very closely related the notions of "authenticity and integrity" of the Chinese conservation principles. As noted by Jin, the intangible factors of architectural heritage in ancient China were specially emphasised. It was particularly due to Taoist philosophy that profoundly influenced the culture of architecture and stressed the inherent relationship of "Being (有)" and "Not-being (无)". This dualism of basic elements, shows the importance of intangible aspects of any material objects in ancient Chinese culture. Therefore, it can be noticed that intangible aspect is always an essential element of building. As a consequence, Chinese architectural heritage transmits not only the physical form of appearance, structure, and materials, but also the intangible elements of aesthetic, lifestyle, handcrafts and cultural connotation beyond its physical appearance. The ultimate goal of any protection practices in a Chinese context is not only the building itself, but spirit and culture being essential part of it. Therefore, Chinese conservation concepts are focused on the preservation of heritage objects in their entirety, thus the preservation of intangible elements is as important as the tangible elements. Without these intangible elements, tangible elements of architectural heritage could be easily changed, demolished or lose their cultural importance (Jin, 2019, p. 93-94). It can be noted, that when highlighting the dichotomy between the physical versus spiritual in the use of the concept of authenticity, as it has place often in internationally accepted notions of authenticity, it is most likely that physical aspect has always been given priority. It is like this because the physical part is what can be seen and can be touched with a specific geographic location (Ndoro, 2018, p. 41). That approach has shaped conservation methodologies bringing them to the believe that they are applicable to any circumstances or cultural aspects, which was the direct consequence of domination of Western theories over the rest of the world. As noticed by Lowenthal, "Europeans rate their own national heritage as so superior it ought to be global" (Lowenthal, 1996, p. 239).

Apart from the underlying the importance of preserving the "whole", The Declaration of Qufu also stresses the necessity of maintaining the "original state" or "original condition" of architectural relict. To be precise the document actually makes an attempt to define a state of a monument's "historic condition". As noted by Zhu, the term "historic condition" is kind of synonymic to the China Principles' notion of "original state", and is described as "a healthy rather than ruinous or ramshackle state. Luo is stressing this notion and is drawing attention to the fact that it is not "the existing state of destruction, distortion, or wear" (Luo, et al., 2005). Therefore, in the event of any alteration in the historic condition, efforts should be made to restore the cultural heritage to its historic condition as early as possible" (Zhu, 2017, p. 191-192). It leads to an observation, that "for damaged architectural ancient buildings, as long as they are carefully repaired according to the original prototype, original materials, original structure and original crafts, their scientific value, artistic value and historical value still exist" (Jin, 2019, p. 95). At this point it is necessary to remember, that the spatial wholeness of Chinese architecture reflects ritual and spirit of the space. Thus, any incompleteness of the architectural structure in consequences brings a weakening of spiritual essence of this structure, as well as the remainder of the site. The overall layout of Chinese architecture, is in parallel the manifestation of intangible elements being inseparable part of the wholeness, such as Taoist and Confucianism philosophy or the Feng Shui concept. It means that any defect resulting of vandalism or natural disaster should be restored to the original condition based on "authentic information". Exceptions can be considered when the defect holds particular commemorative significance or research value (Jin, 2019, p. 95). However, in Article 10 in the *Qufu*

Declaration, it is emphasized that in the case of any reconstruction "the specific analysis and evaluation should be carried out according to the environment, status, the role of the building group, the integrity and authenticity of the damaged architectural relics. The reconstruction of the damaged architectural relics should be approved by experts and the supervisor department of cultural relics" (Luo, et al., 2005). It should be stressed, that following this interpretation the buildings repaired or restored to its historic condition should not be treated thus as "fake antiques". As noticed by Zhu, this particular issue illustrates highly controversial notions of authenticity in China, the understanding of which directly comes from the clash of negotiations on "global" and "local" concerns in heritage conservation (Zhu, 2017, p. 192). But it seems that one another aspect is even more debatable. As noted by Jin, Chinese conservation principles states that the "original condition" somehow requires approach combining of the existing condition and a restored historic condition (Figure 6.1 – 6.26). Therefore, it shows an ambiguous notion of authenticity when comes to preservation of "historic condition" (Jin, 2019, p. 95).

Figure 6.1. The restoration approach of "original condition" as a conglomerate of combining the existing condition and a restored historic condition – in the places where the paint is still preserved it is kept with "time patina" plus some retouching is made imitating "old paint with patina", whereas in the spots where paint is completely lost, it is reconstructed "as it originally was"; Da Gao Xuan Dian Taoist Temple (The Royal Taoist Temple), Beijing; the state during the completion of restoration works in 2018. The temple was built in 1542 during the Ming dynasty and it was originally used by imperial families to worship Taoist deities and wish for sunny days, rain and snow. It had not been renovated for more than a century. The recent complex conservation was undertaken between 2015 to 2018 (photo T. Tomaszek, 2018).

The *Qufu Declaration* particularly prioritizes the methods of conservation of wooden structures that were used traditionally in China. They have developed due to the fact that wooden structures, which consist as a majority of China's architectural heritage, are not easy to sustain for a long period, even with regular preventive maintenance. Therefore, in article 6 of this document is stated, that the one of the most important methods for

Figure 6.2. The restoration approach of "original condition" as a conglomerate of combining the existing condition and a restored historic condition; Da Gao Xuan Dian Taoist Temple (The Royal Taoist Temple), Beijing (photo T. Tomaszek, 2018).

Figure 6.3. The restoration approach of "original condition" as a conglomerate of combining the existing condition and a restored historic condition; Da Gao Xuan Dian Taoist Temple (The Royal Taoist Temple), Beijing (photo T. Tomaszek, 2018).

Figure 6.4. The restoration approach of "original condition" as a conglomerate of combining the existing condition and a restored historic condition; Da Gao Xuan Dian Taoist Temple (The Royal Taoist Temple), Beijing (photo T. Tomaszek, 2018).

Figure 6.5. The restoration approach of "original condition" as a conglomerate of combining the existing condition and a restored historic condition; Da Gao Xuan Dian Taoist Temple (The Royal Taoist Temple), Beijing (photo T. Tomaszek, 2018).

Figure 6.6. The restoration approach of "original condition" as a conglomerate of combining the existing condition and a restored historic condition; Da Gao Xuan Dian Taoist Temple (The Royal Taoist Temple), Beijing (photo T. Tomaszek, 2018).

Figure 6.7. The restoration approach of "original condition" as a conglomerate of combining the existing condition and a restored historic condition; one of the side halls of the Palace of Compassion and Tranquility (Palace built in 1536 and rebuilt 1653, the surrounding structures adapted in 1736), Forbidden City, Beijing (photo T. Tomaszek, 2018).

Figure 6.8. The restoration approach of "original condition" as a conglomerate of combining the existing condition and a restored historic condition; The Palace of Compassion and Tranquility (Cining gong), first built in the Ming Dynasty in 1536, rebuilt in the Qing Dynasty in the year 1653; Forbidden City, Beijing (photo T. Tomaszek, 2018).

Figure 6.9. The restoration approach of "original condition" as a conglomerate of combining the existing condition and a restored historic condition; The Palace of Compassion and Tranquility (Cining gong), first built in the Ming Dynasty in 1536, rebuilt in the Qing Dynasty in the year 1653; Forbidden City, Beijing (photo T. Tomaszek, 2018).

Figure 6.10. The restoration approach of "original condition" as a conglomerate of combining the existing condition and a restored historic condition; The Palace of Compassion and Tranquility (Cining gong), first built in the Ming Dynasty in 1536, rebuilt in the Qing Dynasty in the year 1653; Forbidden City, Beijing (photo T. Tomaszek, 2018).

Figure 6.11. The restoration approach of "original condition" as a conglomerate of combining the existing condition and a restored historic condition; The Palace of Longevity and Health (Shoukang gong), constructed between December 1735 to October 1736; Forbidden City, Beijing (photo T. Tomaszek, 2018).

Figure 6.12. The restoration approach of "original condition" as a conglomerate of combining the existing condition and a restored historic condition; The Palace of Longevity and Health (Shoukang gong), constructed between December 1735 to October 1736; Forbidden City, Beijing (photo T. Tomaszek, 2018).

Figure 6.13. The restoration approach of "original condition" as a conglomerate of combining the existing condition and a restored historic condition; The Palace of Longevity and Health (Shoukang gong), constructed between December 1735 to October 1736; Forbidden City, Beijing (photo T. Tomaszek, 2018).

Figure 6.14. The restoration approach of "original condition" as a conglomerate of combining the existing condition and a restored historic condition; West side gate to the courtyard between The Palace of Heavenly Purity and the Gate of Heavenly Purity; Forbidden City, Beijing (photo T. Tomaszek, 2018).

Figure 6.15. The restoration approach of "original condition" as a conglomerate of combining the existing condition and a restored historic condition; West side gate to the courtyard between The Palace of Heavenly Purity and the Gate of Heavenly Purity; Forbidden City, Beijing (photo T. Tomaszek, 2018).

Figure 6.16. The restoration approach of "original condition" as a conglomerate of combining the existing condition and a restored historic condition; Xianruo Hall in the garden of the Hall of Consolation of Mothers (The Garden of Benevolent Tranquility); Forbidden City, Beijing (photo T. Tomaszek, 2018).

Figure 6.17. The restoration approach of "original condition" as a conglomerate of combining the existing condition and a restored historic condition; Xianruo Hall in the garden of the Hall of Consolation of Mothers (The Garden of Benevolent Tranquility); Forbidden City, Beijing (photo T. Tomaszek, 2018).

Figure 6.18. The restoration approach of "original condition" as a conglomerate of combining the existing condition and a restored historic condition; the auxiliary hall of Xianruo Hall in the garden of the Hall of Consolation of Mothers (The Garden of Benevolent Tranquility); Forbidden City, Beijing (photo T. Tomaszek, 2018).

Figure 6.19. The restoration approach of "original condition" as a conglomerate of combining the existing condition and a restored historic condition; the Drum Tower (erected in 1380 during the early Ming Dynasty) in Xi`an, Shaanxi Province (photo T. Tomaszek, 2018).

Figure 6.20. The restoration approach of "original condition" as a conglomerate of combining the existing condition and a restored historic condition; the Drum Tower (erected in 1380 during the early Ming Dynasty) in Xi`an, Shaanxi Province (photo T. Tomaszek, 2018).

Figure 6.21. The restoration approach of "original condition" as a conglomerate of combining the existing condition and a restored historic condition; The Stele Pavilion, the Confucian Temple (built in 1302-1306), Beijing (photo T. Tomaszek, 2018).

Figure 6.22. The restoration approach of "original condition" as a conglomerate of combining the existing condition and a restored historic condition; The Stele Pavilion, the Confucian Temple (built in 1302-1306), Beijing (photo T. Tomaszek, 2018).

Figure 6.23. The restoration approach of "original condition" as a conglomerate of combining the existing condition and a restored historic condition; Gate Tower (built in Ming Dynasty) at the Barbican of the West Gate of Xi`an city wall (Anding Gate, originally named Shunyi Gate), Xi`an, Shaanxi Province (photo T. Tomaszek, 2018).

Figure 6.24. The restoration approach of "original condition" as a conglomerate of combining the existing condition and a restored historic condition; Gate Tower (built in Ming Dynasty) at the Barbican of the West Gate of Xi`an city wall (Anding Gate, originally named Shunyi Gate), Xi`an, Shaanxi Province (photo T. Tomaszek, 2018).

Figure 6.25. The restoration approach of "original condition" as a conglomerate of combining the existing condition and a restored historic condition; The Hall of Character Cultivation, Forbidden City, Beijing (photo T. Tomaszek, 2018).

Figure 6.26. The restoration approach of "original condition" as a conglomerate of combining the existing condition and a restored historic condition; The Hall of Character Cultivation, Forbidden City, Beijing (photo T. Tomaszek, 2018).

restoration is dismantling followed by disassembling, which is a thorough and effective traditional repair methods for architectural relics (Luo, et al., 2005). Despite the advanced interference of historical fabric this kind of metabolic protection method still conforms to the principle of authenticity. This is particularly applicable in the case of religious structures, where renewal of components and exchange of decaying materials are necessary for sustaining continuity of a sacred place. The process of dismantling and reassembly, feasible due to special non-rigid joinery Sunmao (榫卯) making buildings flexible and at the same time easy to disassemble, also allows the ancient architectural crafts and techniques to be maintained whilst respecting historical information. Therefore, it allows us also to increase our understanding of ancient construction methods. Knowledge of which is essential for their proper preservation. The significance of ancient construction lies on not only the existence of a specific building in a certain historical period, but to the same importance is the continuous renewal and circulation of specific architectural forms (Figure 6.27 – 6.36). Thus, the process of continuous reconstruction of China's ancient construction makes natural metabolism and continuation, which is an essential way for the knowledge of wooden frame structures to be handed down to future generations (Jin, 2019, p. 96-99).

The *Qufu Declaration* also includes a recommendation on architectural painting, especially wood components painting. In Article 7 of the document, it is stressed that the painting is meant "to protect the wood and beautify the building. At the same time, it also represents the architectural level and the function of the building. Therefore, if the remaining architecture paint layer has completely lost the original function and it is of no artistic value anymore, it should be re-painted" (Luo, et al., 2005). It must be remembered, that the architectural painting serves in traditional wooden building in China not only as an artistic decoration, but is also applied to protect the wood from any chemical or physical factors. Due to the effects of weathering, the layer of paint cannot remain intact over a long period of time. Thus, it is very important for preservation of the wooded materials used in

Figure 6.27. Reconstruction of the back part of Hall of Sakyamuni, The Dailuoding Temple (built in the Ming Dynasty, repaired and reconstructed in the following Ming and Qing Dynasties), Wutai Shan (Mount Wutai, also known as Mount Qingliang), one of the four sacred mountains of Buddhism in China, Shanxi Province (photo T. Tomaszek, 2019).

Figure 6.28. Reconstruction of the back part of the Hall of Sakyamuni, The Dailuoding Temple (built in the Ming Dynasty, repaired and reconstructed in the following Ming and Qing Dynasties), Wutai Shan (Mount Wutai, also known as Mount Qingliang), one of the four sacred mountains of Buddhism in China, Shanxi Province (photo T. Tomaszek, 2019).

Figure 6.29. Reconstructions of wooden temple décor using traditional crafts; The Hall of the Buddha of Wealth, The Guangren Lama Temple (Tibetan Buddhist Monastery built in 1703), Xi`an, Shaanxi Province. It is the only temple of the Gelug School of Tibetan Buddhism in Shaanxi Province and it is used to serve as a Palace Hotel for Dalai Lamas (photo T. Tomaszek, 2018).

Figure 6.30. Reconstruction of wooden temple décor using traditional crafts; The Hall of the Buddha of Wealth, The Guangren Lama Temple (Tibetan Buddhist Monastery built in 1703), Xi`an, Shaanxi Province. It is the only temple of the Gelug School of Tibetan Buddhism in Shaanxi Province and it is used to serve as a Palace Hotel for Dalai Lamas (photo T. Tomaszek, 2018).

Figure 6.31. A traditional tool used for reconstruction of wooden temple décor; The Hall of the Buddha of Wealth, The Guangren Lama Temple (Tibetan Buddhist Monastery built in 1703), Xi`an, Shaanxi Province (photo T. Tomaszek, 2018).

Figure 6.32. The living tradition of traditional building still present in the remote Tongle village (near Yezhi), Yunnan Province. The entire village consists of only traditional log structures which are maintained in the traditional way (photo T. Tomaszek, 2019).

Figure 6.33. The living tradition of traditional building still present in the remote Tongle village (near Yezhi), Yunnan Province. The entire village consists of traditional log structures which are restored in the traditional way (photo T. Tomaszek, 2019).

Figure 6.34. The living tradition of traditional building still present in the remote Tongle village (near Yezhi), Yunnan Province. The entire village consists of traditional log structures which are restored in the traditional way (photo T. Tomaszek, 2019).

Figure 6.35. Reconstruction of a traditional wooden dwelling; Lige Village on the shore of Lugu Lake, Yunnan Province (photo T. Tomaszek, 2019).

Figure 6.36. Marks on the logs made before dismantling the log dwelling; Lige Village on the shore of Lugu Lake, Yunnan Province (photo T. Tomaszek, 2019).

building construction to be re-painted at regular intervals (Figure 6.37 – 6.69). The basic principle of re-painting should follow the "Four Originals", which refers to the authenticity of the structure's historic state of original prototype, original materials, original structure and original crafts (Figure 6.29 – 6.31, 6.47 – 6.69). Repainting should also be developed with proper professional consultation (Jin, 2019, p. 98-99). In Article 31 of the *Principles for the Conservation of Heritage Sites in China,* it is underlined that it is not permissible to redo decorative painting for new or cosmetic purposes, or for any further development. Any re-painting should be performed only for the needs of restoration, which can not result in falsification of the original cultural and societal value of the site (ICOMOS China, 2015 (a)).

Figure 6.37. Re-painted on regular basis the surface of elements directly exposed to external atmospheric factors, thus degrading at an accelerated pace – traditional preservation methods are meant to protect the wood and beautify the building, combining of the existing condition and a restored condition; Da Cheng Gate (The Gate of Great Success), Confucian Temple (KongMiao) built in 1302-1306, Beijing (photo T. Tomaszek, 2018).

Following the proclamation of The *Qufu Declaration* there were continuous disputes focusing on conservation principles applicable in the eastern world. In the years following, numerous international conferences were held in China dedicated to architectural heritage conservation in the context of the so-called "Orient". Among the most important events, an International Symposium on the Concepts and Practices of Conservation and Restoration of Historic Buildings in East Asia was held in Beijing in 2007. The event was organized by SACH (State Administration of Cultural Heritage in China), ICCROM (International Centre for the Study of the Preservation and Restoration of Cultural Property), ICOMOS and the UNESCO World Heritage Centre. The meeting was dedicated to developing recommendations for the 30th Congress of the World Heritage Committee, where conservation methodologies applied on the most important heritage objects in China – the Forbidden City, the Temple of Heaven and the Summer Palace in Beijing were discussed. The findings have been summarized in the document "*Beijing Document on the Conservation and Restoration of Historic Buildings in East Asia*'. One year later, in 2008, the

Figure 6.38. Re-painted on regular basis the surface of elements directly exposed to external atmospheric factors, thus degrading at an accelerated pace – traditional preservation methods are meant to protect the wood and beautify the building, combining of the existing condition and a restored condition; The Gate of Supreme Harmony, the Forbidden City, Beijing (It was originally built during the Ming Dynasty, then it burned down in 1886 due to a fire started by a broken lamp in the guard room. The present gate dates from the rebuilding after this fire, which was completed in 1894); (photo T. Tomaszek, 2018).

Figure 6.39. Re-painted on regular basis the surface of elements directly exposed to external atmospheric factors, thus degrading at an accelerated pace – traditional preservation methods are meant to protect the wood and beautify the building, combining the existing condition and a restored condition; The Gate of Manifest Virtue (Zhaode men) built in the Ming Dynasty in 1420, destroyed in a fire during the Guangxu reign in 1888, and rebuilt in the following year, the Forbidden City, Beijing (photo T. Tomaszek, 2018).

Figure 6.40. Re-painted on regular basis the surface of elements directly exposed to external atmospheric factors – traditional preservation methods are meant to protect the wood and beautify the building; The Hall of Supreme Harmony (TaiHe Dian), commonly known as the Hall of Golden Bells (Jinluan Dian), built in the Ming dynasty in 1420; the Forbidden City, Beijing (photo T. Tomaszek, 2018).

Figure 6.41. Re-painted on regular basis the surface of elements directly exposed to external atmospheric factors– traditional preservation methods are meant to protect the wood and beautify the building; The Palace of Compassion and Tranquility (Cining gong), built in the Ming Dynasty in 1536 and rebuilt in the Qing Dynasty in 1653, and later in 1736; the Forbidden City, Beijing (photo T. Tomaszek, 2018).

Figure 6.42. Re-painted on regular basis the surface of elements directly exposed to external atmospheric factors, thus degrading at an accelerated pace – traditional preservation methods are meant to protect the wood and beautify the building, combining of the existing condition and a restored condition; The Palace of Compassion and Tranquility (Cining gong), the Forbidden City, Beijing (photo T. Tomaszek, 2018).

Figure 6.43. Re-painted on regular basis the surface of elements directly exposed to external atmospheric factors– traditional preservation methods are meant to protect the wood and beautify the building; The Big Buddha Hall north of The Palace of Benevolent Tranquility; the Forbidden City, Beijing (photo T. Tomaszek, 2018).

Figure 6.44. Re-painted on regular basis the surface of elements directly exposed to external atmospheric factors– traditional preservation methods are meant to protect the wood and beautify the building; The Palace of Longevity and Health (Shoukang gong), constructed between December 1735 to October 1736, the Forbidden City, Beijing (photo T. Tomaszek, 2018).

Figure 6.45. Exterior of the Hall of Maitreya after removing the remains of degraded layer of paint, before re-painting; the Wuyou Buddhist Temple (formerly Zhengjue Temple founded during Zhide period and Qianyuan period of the Tang dynasty) located on the top of Mount Wuyou, Shizhong District of Leshan, Sichuan Province. The temple was destroyed and rebuilt many times. The present version was completed in the Qing dynasty (1644–1911). It has been designated as a Key National Buddhist Temple in Han Chinese Area by the State Council of China in 1983 (photo T. Tomaszek, 2018).

Figure 6.46. Exterior of the Hall of Maitreya after removing the remains of degraded layer of paint, before re-painting; Wuyou Buddhist Temple, Shizhong District of Leshan, Sichuan Province (photo T. Tomaszek, 2018).

Figure 6.47. Re-painting works in the Arhat Hall; Wuyou Buddhist Temple, Shizhong District of Leshan, Sichuan Province (photo T. Tomaszek, 2018).

Figure 6.48. Re-painting works in the Arhat Hall; Wuyou Buddhist Temple, Shizhong District of Leshan, Sichuan Province (photo T. Tomaszek, 2018).

Figure 6.49. Hall of Guanyin after re-painting the exterior; Wuyou Buddhist Temple, Shizhong District of Leshan, Sichuan Province (photo T. Tomaszek, 2018).

Figure 6.50. Re-painting work on the elements with degraded layer of paint (complete removal of totally degraded paint and re-painting from scratch using traditional technology); San Xi Tang (Room of Three Rarities), Hall of Mental Cultivation (Yangxin dian), built in 1746; Forbidden City, Beijing (photo T. Tomaszek, 2018).

Figure 6.51. Re-painting work on the elements with degraded layer of paint (complete removal of totally degraded paint and re-painting from scratch using traditional technology); San Xi Tang (Room of Three Rarities), Hall of Mental Cultivation (Yangxin dian), built in 1746; Forbidden City, Beijing (photo T. Tomaszek, 2018).

Figure 6.52. Re-painting work on the elements with degraded paint (complete removal of totally degraded paint and re-painting from scratch using traditional technology); San Xi Tang (Room of Three Rarities), Hall of Mental Cultivation (Yangxin dian), built in 1746; Forbidden City, Beijing (photo T. Tomaszek, 2018).

Figure 6.53. Door re-painting due to a degraded paint layer (complete removal of totally degraded paint and all technological layers under the paint layer, re-painting from scratch using traditional technology and traditional materials for preparing all the layers); The doors of the side entrances in the area of the Six Western Palaces in the Inner Court, Forbidden City, Beijing (photo T. Tomaszek, 2018).

Figure 6.54. Door re-painting due to a degraded paint layer (complete removal of totally degraded paint and all technological layers under the paint layer, re-painting from scratch using traditional technology and traditional materials for preparing all the layers); The doors of the side entrances in the area of the Six Western Palaces in the Inner Court, Forbidden City, Beijing (photo T. Tomaszek, 2018).

Figure 6.55. Conservation works – removing the degraded paint layer together with all technological layers under it from the lower parts of the columns/pillars/windows and lower part of the building (April 2018); The Palace of Tranquil Longevity (Ningshou gong) built in the Ming Dynasty and originally named Renshou gong (Palace of Benevolent Longevity). After its repair in 1689 it was renamed the rear hall of the Palace of Tranquil Longevity. In 1776 it was reconstructed in imitation of the Palace of Earthy tranquility (Kunning gong), Forbidden City, Beijing (photo T. Tomaszek, 2018).

Figure 6.56. Re-painted surfaces using traditional technology and traditional materials; (July 2018); The Palace of Tranquil Longevity (Ningshou gong), Forbidden City, Beijing (photo T. Tomaszek, 2018).

Figure 6.57. Conservation works – removing the degraded paint layer together with all technological layers under it from the lower parts of the columns/pillars/windows and lower part of the building (April 2018); The Palace of Tranquil Longevity (Ningshou gong), Forbidden City, Beijing (photo T. Tomaszek, 2018).

Figure 6.58. Re-painted surfaces using traditional technology and traditional materials; (July 2018); The Palace of Tranquil Longevity (Ningshou gong), Forbidden City, Beijing (photo T. Tomaszek, 2018).

Figure 6.59. Conservation works – the removal of degraded paint together with all technological layers under it from the lower parts of the columns/pillars/windows and lower part of the building (April 2018); The Palace of Tranquil Longevity (Ningshou gong), Forbidden City, Beijing (photo T. Tomaszek, 2018).

Figure 6.60. Re-painted surfaces using traditional technology and traditional materials (July 2018); The Palace of Tranquil Longevity (Ningshou gong), Forbidden City, Beijing (photo T. Tomaszek, 2018).

Figure 6.61. Conservation works – removal of degraded paint together with all technological layers under it from the lower parts of the columns/pillars/windows and lower part of the building (April 2018); The Palace of Tranquil Longevity (Ningshou gong), Forbidden City, Beijing (photo T. Tomaszek, 2018).

Figure 6.62. Re-painted surfaces using traditional technology and traditional materials (July 2018); The Palace of Tranquil Longevity (Ningshou gong), Forbidden City, Beijing (photo T. Tomaszek, 2018).

Figure 6.63. Conservation works – removal of degraded paint together with all technological layers under it from the lower parts of the columns/pillars/windows and lower part of the building (April 2018); The Palace of Tranquil Longevity (Ningshou gong), Forbidden City, Beijing (photo T. Tomaszek, 2018).

Figure 6.64. Re-painted surfaces using traditional technology and traditional materials (July 2018); The Palace of Tranquil Longevity (Ningshou gong), Forbidden City, Beijing (photo T. Tomaszek, 2018).

Figure 6.65. Conservation works – removal of degraded paint together with all technological layers under it (April 2018); The Hall of Imperial Supremacy (Huangji dian), built around 1771, the main hall in the area of Palace of Tranquil Longevity (Ningshou gong), which the Qianlong Emperor (r. 1736-1795) built for spending his years after his retirement, Forbidden City, Beijing (photo T. Tomaszek, 2018).

Figure 6.66. Re-painted surfaces using traditional technology and traditional materials (July 2018); The Hall of Imperial Supremacy (Huangji dian), Forbidden City, Beijing (photo T. Tomaszek, 2018).

Figure 6.67. Re-painted on regular basis the exterior surface of the temple – traditional preservation methods are meant to protect the wood and beautify the building; The Guangren Lama Temple (Tibetan Buddhist Monastery built in 1703), Xi`an, Shaanxi Province. It is the only temple of the Gelug School of Tibetan Buddhism in Shaanxi Province and it used to serve as a Palace Hotel for Dalai Lamas (photo T. Tomaszek, 2018).

Figure 6.68. Re-painted on regular basis the exterior surface of the temple – traditional preservation methods are meant to protect the wood and beautify the building; The Central Hall, The Guangren Lama Temple (Tibetan Buddhist Monastery built in 1703), Xi`an, Shaanxi Province (photo T. Tomaszek, 2018).

Figure 6.69. Re-painted on regular basis the exterior surface of the temple – traditional preservation methods are meant to protect the wood and beautify the building, The Guangren Lama Temple (Tibetan Buddhist Monastery built in 1703), Xi`an, Shaanxi Province (photo T. Tomaszek, 2018).

next significant international meeting was held – the International Symposium on the Conservation of Painted Wood Architectural Surfaces in East Asia. After the discussions, the *"Beijing Memorandum on the Conservation and Restoration of Caihua (decorative painting on wood) in East Asia"* was adopted. All of these new charters and ongoing, complex disputes were the reason for revising once more the *Principles of the Conservation for Heritage sites in China.* The newest edition was finally published in 2015. The document stresses the importance of protection methods with Chinese characteristics on architectural heritage, and provides a detailed recommendation regarding the issues of replacement, reconstruction, authenticity, and architectural painting methods (Jin, 2019, p. 92).

Living Heritage and continuity as a reference level for authenticity in a Chinese context

ABSTRACT: The stress placed on the special approach towards conservation of living monuments in contemporary conservation theories, has brought to attention the issue of 'continuity' that should be considered as a key feature that characterizes living heritage. After the concept of 'Living Heritage' was introduced to conservation theory and practice of Asian countries, it has significantly influenced the understanding of wooden heritage protection in this part of the World. As noted by some researchers, the traditional Asian approach is diametrically opposed to western, particularly European approaches. While the first is more about the understanding of a culture that has been going through a continuous process of adaptation and change, the second, in contrary, rather than looking at changes in the structure over time, is fixed upon very specific changes in specific time periods. This chapter presents the development of the idea of "living heritage" and its understanding in China, as well as discussing the issue of continuity and repetition pointing at the fact, that heritage sites do not remain static but are in constant change. Subsequently, the notion of authenticity regarded as the "management of change and continuity" is discussed.

In search for the notion of authenticity in the preservation of wooden architecture in China, it is quite necessary to look at the protection of the country's cultural heritage from the perspective of the recent decades and the changes related to it. China is a country of enormous contrasts observable in every area of life, resulting from the vast area and various geographical, as well as cultural conditions. China is also a land where, apart from the dominant Han community, several dozen ethnic minorities inhabit. All of this has influenced the formation of a number of different architectural traditions and ways of capturing and interpreting the surrounding reality.

It is remarkable that after adopting a socialist system, China has largely distanced itself from traditional religious systems. Nevertheless, places and traditions with a special dimension to the cultural heritage of this country have survived there, where wooden architecture is probably the most important form of manifestation of local identity. These are the Sacred Mountains, lands of the ethnic minorities and the Tibet Autonomous Region. Historical architectural structures present in these places, among which the most prominent, are temples and other sacred buildings. Which are still used for age-old practices and materialize the cosmological and axiological systems of their users. The growing awareness of the heritage of their own country, and thus, the need to protect historical architecture in today's China, posed a difficult question as to the appropriate approach to the protection of these places. The conglomerate of domestic conservation philosophies and those borrowed from international legislation presented many new challenges. The search for understanding of authenticity and the possibility of maintaining it as a superior value draws attention to the fact of the living traditions of these places. Revealed through the continuation of eternal cultural systems manifested by means of a strictly defined material form. At the same time, the issues so important for these areas reflect the nature of the broadly understood Asian culture, whose experience in the conservation of cultural heritage has contributed to the dissemination of the concept of living heritage.

The concept of 'Living Heritage' has been recently developed as an important voice within heritage discourse. As noted by Wijesuriya, it comes from debates on living vs. dead monuments that originated in the formative period of modern conservation theory (Wijesuriya, 2018, p. 47). A particularly important document dedicated to this issue was *Recommendations of the Madrid Conference* adopted in 1904. It proposed to divide monuments into two classes: [1] *dead monuments* – those belonging to a past civilization or serving obsolete purposes, and [2] *living monuments* – those which continue to serve the purposes for which they were originally intended (Recommendations of the Madrid Conference…, 1909).

Subsequently, there are some resolutions regarding the protection of *living monuments*:

- They "ought to be *restored* so that they may continue to be of use, for in architecture utility is one of the bases of beauty".
- "Such restoration should be effected in the original style of the monument, so that it may preserve its unity, unity of style being also one of the bases of beauty in architecture, and primitive geometrical forms being perfectly reproducible. Portions executed in a different style from that of the whole should be respected, if this style has intrinsic merit and does not destroy the aesthetic balance of the monument" (Recommendations of the Madrid Conference…, 1909).

The important notion of particular approach to *living monuments* is also described by Charles Peers, who noted: "There is a great distinction between buildings which are still occupied and buildings which are in ruins. Buildings which are in use are still adding to their history; they are alive. […] When you come to a building which is being used as a dwelling house, or a church…[…] You have to perpetuate it as a living building, one adapted to the use of the present generation, but which has a history to be preserved" (Forsyth, 1913, p. 135).

The concept of *living monuments*, developed in the western civilization, has become particularly important when introduced in other parts of the World, especially Asian countries. In 1923, during the field works in India John Marshal suggested, that when we deal with living monuments sometimes they require to restore them to a greater extent than it is desirable on purely archaeological sites (Marshall, 1990, paragraphs 25, 26). In Sri Lanka Paranavitane in 1945 he wrote "restoration of ancient shrines […] has to be carried out without hurting the religious susceptibilities of the people" (Paranavitana, 1945, p. 43).

The stress placed on the special approach towards conservation of *living monuments* has brought to attention the issue of the 'use' of heritage building or the 'purposes for which it was originally intended'. It is because it links directly to the 'living' dimension of heritage. Therefore, transferring it from the built heritage to any kind of heritage, as noted by Wijesuriya, the term 'Living Heritage' is naturally connected to 'communities' and the 'continuity' of traditions and practices. And it is 'continuity' that should be considered as a key feature that characterize living heritage, whereas 'use for which it was originally intended' is the essential factor of defining 'continuity'. It should be noted, however, that this 'use' required for 'continuity' should not be confused with the notion of 'use' as some form of functioning or use that heritage sites have for the present society. The 'use' referred to here is an idea that determines the authentic purpose of the building, and thus can be treated as determining the scope of authenticity when evaluating a given object (Wijesuriya, 2018, p. 47).

Trying to define the living heritage, it could be described that it is characterized by "the continuity of the original function or the purpose for which it was originally established. […] such heritage maintains the continuity of community connections, continues to evolve in the form of tangible and intangible expressions and taken care through traditional or established means" (Wijesuriya, 2018, p. 51-52). As noted by Wijesuriya, it is possible to identify three elements of continuity: [1] Continuity of community connections; [2] Continuity of cultural expressions (both tangible and intangible); [3] Continuity of care (through traditional or established means) (Wijesuriya, 2018, p. 50). Continuity of the original function into the present comes in general from association or connection to a certain community for whom particular heritage site was created, and which life is influenced by and influences such heritage. Therefore, this certain community should have the deciding voice within the decision making process in preservation decisions, this is confirmed in the Nara Document: "responsibility for cultural heritage and the management of it belongs, in the first place, to the cultural community that has generated it, and subsequently, to that which cares for it". It needs to be noted however, that the mentioned 'community' may actually consist group of people, but in some cases may also include only some individuals. Therefore, the basic issue is the accurate identification of a community (Wijesuriya, 2018, p. 50).

It needs to be noted that in the last decades the definition of heritage has been expanded to include not only architectural structures or historical objects, but also complex living sites together with all human activities and changing uses and significance (Weise, 2018, p. 121). Therefore, it is justified to postulate that the conservation of wooden heritage of living communities must guarantee: "the continuation of a living cultural heritage which must be based on the appreciation of the past, however, taking change into account" (the Department of Archaeology, 2007, p. 5). Whenever continuity of a living culture exists within a particular physical setting (for example in the case of a wooden temple or a sacred landscape together with sacred trees and wooden structures), what should be given priority is nothing else but continuity. Continuity then requires particular processes of adaptation, however, this adaptation must to be appropriate to the given context. The wooden living heritage, that is still being used or worked on, goes therefore through cyclical change which is an aspect of continuity, and this should be considered as the appropriate approach for the preservation of this heritage. Thus, an understanding of continuity is essential, as continuity might be of use, of physical structure, or even significance (Weise, 2018, p. 130), and it could attribute in various ways of materialization.

The tradition of periodic reconstruction of historic wooden buildings has been present in Asia since ancient times. Particularly, this tradition is strongly rooted in Japan, where it is common to observe the process of continually maintaining religious structures by replacing deteriorated parts or periodically tearing down and reconstructing entire structures. This process simultaneously acts as a guarantee of continually maintaining the common bond of the community, thus accepting any natural changes in the rituals of the congregation. It also allows the possibility of upgrading and adapting the temple to new requirements.

As noted by Weise, this traditional Asian approach is diametrically opposed to the western, particularly European approach. While the first is more about the understanding of a culture that has been going through a continuous process of adaptation and change, the second one, in contrary, rather than looking at changes in the structure over time is fixed upon very specific changes in specific time periods. And this fixed time is treated as a baseline for preservation in western tradition, but continuity naturally included in tangible as well as intangible aspects of cultural heritage does not necessarily provide justification for the fixed time in question (Weise, 2018, p. 125).

There is an important issue that comes to mind when talking about continuity and heritage. It seems that one of the main reasons for modern conservation philosophies, considering heritage building as freezing in a given space and time was an understanding that the historical continuity between the past and present was already broken. The process of continuity was therefore overlooked or even ignored. It is well underlined by Philippot, who noticed that the past was considered in Western culture to be already completed, now it is observed from a distance and seen as a "panorama" (Philippot, 1996, p. 268). Mentioned above, the introduction of the concept of *living monuments* in a non-western context has brought some critical observations. One of them was to accept that the links between the past, present and future does not need to be considered as linear and that they are treated differently in various societies. As noted by Anyon, the concept of the protection of the past seems to be quite simple. However, the nature of "past" as well as the nature of its "protection" are defined depending on the culture (Anyon, 1991, p. 218). This is the case for Native-Americans, for whom time is not a linear concept. For some, like the Zuni tribe, the present "does not have to look like the past because the past lives on in the everyday actions [...] The essential cultural difference is that non-Indians want to see the past to know it, whereas to Indians the present embodies the past and thus they do not necessarily have to see their past to know it" (Anyon, 1991, p. 220). Also, according to Matunga, for Maori People past belongs to the "living present". The same there is no a firm line between the past and the present, and thus the past is not relinquished in favor of the present (Matunga, 1994, p. 219). A very special understanding of time, comes from Hindu tradition. in Hinduism, time is considered as a cyclical concept, and thus the process of creation moves in cycles and it is never ending. This is also the true for Buddhism, which has adopted the concept of samsara, the wheel of life. Where everything occurs in cycles, repeating the natural order, where everything "begins to end and ends to begin" (Wijesuriya, 2010, p. 240).

It should be noted that, alongside Taoism and Confucianism, it is the Buddhist tradition that has played a fundamental role in shaping Chinese culture. Thus, the concept of the cyclical nature of time and the natural process of repetition are permanently present in the country's intangible heritage. At the same time, it is the Buddhist temples that occupy a special place among the most impressive Chinese historic wooden structures. The vast majority of them, especially those located in the Sacred Mountains or in the Tibet Autonomous Region are one of the most outstanding examples of the importance of living heritage in China.

It is worth noting, that cyclicality and repetition is what reflects the natural character of wooden architectural traditions in Asian countries. This aspect is of fundamental importance in trying to capture the problem of authenticity in the protection of the architectural heritage of these countries, particularly China. It is especially here that traditional wooden architecture materializes a strictly established and repeated for centuries "pattern", where "continuity" is somehow inscribed in its genotype. At the same time, this "continuity" manifests itself not only in the constant transfer of the established form, detail or meaning of colors, but also in transmitting intangible values that determine the aforementioned aspects. Such a "continuity" of the architectural structure, transmitted over time in a fixed form, naturally stands in opposition to the permanent progression of formal development, observed in the architectural tradition of Western cultures.

Interestingly, the natural process of continuity and repetition of patterns, broken during the Cultural Revolution, manifests itself in an increased need to replicate historical buildings in contemporary China. At the same time, until recently disregarded, the architectural heritage of ethnic minorities is now of particular interest, as many of the processes of cultural continuity has, as yet often not been disrupted (Figure 7.1 – 7.8). This recognition of the importance of its own heritage is an attempt for China to rebuild the broken link of continuity. This process also brings a direct response to the attempt of defining the semantic scope of the notion of authenticity, a foreign term, and a little automatically transferred from international conservation legislation. As a value in itself, authenticity relying on continuity, may shift the meaning of preserving China's wooden architectural heritage from a preservation perspective focused on fabric to a value-focused approach. However, this type of static model can only not fully describe the tradition of continuity of architectural form in China, but neither the concept of living heritage as a continuity of the original purpose of the building.

From a living heritage viewpoint, it becomes obvious that the heritage of place does not remain static but is in constant change. As noted by Wijesuriya, change is natural process to facilitate the function of place. This change applies to various tangible or intangible

Figure 7.1. Traditional "Ganlan" style wooden dwellings, Longsheng Dazhai Red Yao Minority Village, Jinkeng Rice Terraced Fields (part of Longji Rice Terraces), Longsheng County, Guangxi (formerly a province, an autonomous region from 1958); (photo T. Tomaszek, 2018).

Figure 7.2. Traditional "Ganlan" style wooden dwellings, Longsheng Dazhai Red Yao Minority Village, Jinkeng Rice Terraced Fields (part of Longji Rice Terraces), Longsheng County, Guangxi (formerly a province, an autonomous region from 1958); (photo T. Tomaszek, 2018).

Figure 7.3. Traditional "Ganlan" style wooden dwellings, Longsheng Dazhai Red Yao Minority Village, Jinkeng Rice Terraced Fields (part of Longji Rice Terraces), Longsheng County, Guangxi (formerly a province, an autonomous region from 1958); (photo T. Tomaszek, 2018).

Figure 7.4. Traditional wooden structures at Zhaoxing Dong Village, Guizhou Province. The village, belonging to the ethnic minority of Dong, was built in 1160A.D. (in the 5th year of the Southern Song Dynasty). It is the most famous of the group of eight Dong Villages, called sometimes by similar name as "Zhaoxing Dong Villages" (photo T. Tomaszek, 2018).

Figure 7.5. Traditional wooden structures at Zhaoxing Dong Village, Guizhou Province. The village, belonging to the ethnic minority of Dong, was built in 1160A.D. (in the 5th year of the Southern Song Dynasty). It is the most famous of the group of eight Dong Villages, called sometimes by similar name as "Zhaoxing Dong Villages" (photo T. Tomaszek, 2018).

Figure 7.6. Traditional wooden dwellings at Zhaoxing Dong Village, Guizhou Province. The village, belonging to the ethnic minority of Dong, was built in 1160A.D. (in the 5th year of the Southern Song Dynasty). It is the most famous of the group of eight Dong Villages, called sometimes by similar name as "Zhaoxing Dong Villages" (photo T. Tomaszek, 2018).

Figure 7.7. Traditional log dwellings, Da Luoshui Village on the Lugu lakeshore, Yunnan Province. This village is a home to the Mosuo minority ethnic group famous for matriarchal society and their wooden log architecture. Lugu lake, an alpine lake at an elevation of 2,685 metres, located on the border of Yanyuan County, Sichuan Province and Ningland Yi Autonomous County, Yunan Province. The lake`s shores are inhabited by many ethnic minority groups, such as the Mosuo, Norzu, Yi, Pumi and Tibetan. The most numerous of these are the Mosuo people (photo T. Tomaszek, 2019).

Figure 7.8. Traditional log dwellings, Lige Village on the Lugu lakeshore, Yunnan Province. Similarly, as Da Luoshui village, Lige village is a home to Mosuo minority ethnic group (photo T. Tomaszek, 2019).

expressions, but also to the existing tangible and intangible components. Thus, the change may be recognized as an inevitable phenomenon, for which reason heritage wooden buildings should not just be treated as "a physical object" ultimately settled in time and space. Therefore, the preservation of material manifestations, becomes only a factor of maintaining continuity, and consequently keeping the value of authenticity. Whereas, the aspect of change denotes an inherited intangible meaning of physical dimensions. However, it needs to be stressed, that continuity directly comes from community-developed (within their own particular definitions) knowledge systems for any kind of maintenance, interventions, extensions and renewal of buildings and their overall management (Wijesuriya, 2018, p. 50).

Since beginning of 21st century, a turning point in increased significance of heritage conservation in China, cultural landscapes and living heritage have also become very important parts of conservation practice (Zhou, 2018, p. 96). Coming from 2015, the updated version of the China Principles emphasized, apart from the tangible element, but also long-established cultural tradition. This reflects a new understanding in China of the importance of an integrated idea of heritage and its conservation, which is an essential improvement in the conservation practice for Chinese heritage professionals (Zhou, 2018, p. 99). Article 13 and associated commentary of the aforementioned updated China Principles, refers to cultural traditions as follows:

"When a heritage site's values depend on the continuation of associated cultural traditions, consideration needs to be given to preserving these traditions along with the site itself. Conservation of a heritage site may also involve conservation of cultural diversity. Sites may be places where traditional activities are still practiced or may be associated with special ways of production, life-styles or other intangible cultural heritage. These traditions, means of production, ways of life and intangible cultural heritage are important values of a site. Their continuation should be encouraged. Conservation of heritage sites should enable these types of traditional activities, production, ways of life and intangible heritage to adapt to the needs of contemporary life while maintaining their vitality" (ICOMOS China, 2015 (b), p. 71).

Since the official Chinese legislation changed to state, that authenticity must be looked through and regarded- not only for example in any historic district as was common before, but also in any cultural, spiritual or social information of a place and thus, it must be protected. At the same time, it was postulated that the interpretation of authenticity was an important issue for consideration and must be suitably linked to display and objectively explain the values of a given place (Zhou, 2018, p. 99). It was expressed in more detail in the China Principles from 2015, where it is clearly stated:

"Sites that retain their original historic function, particularly those where the traditional way of life has become an integral part of the site's values should be encouraged to continue that function. Sites such as historically and culturally famous cities, towns and villages, as well as cultural landscapes, having developed modern ways of life, while continuing their historic function, demonstrate a distinct cultural significance, and are living heritage. This heritage should continue its historic function while conserving traditional activities and ways of life. Ensuring continuing historic function is a means of conserving the values of this heritage. When managing such a site, special effort should be made to protect the original function. Changes to the use should only be considered after careful consideration. Special attention should be given to avoid the transformation of a residential precinct into a commercial district, as this seriously diminishes its values and authenticity. When living sites undertake necessary repairs, adjustments and modifications to satisfy modern needs care must be taken to ensure that there is no damage or alteration to a site's attributes and values. It is not permissible to change historic forms, structure, techniques, materials, decoration and the setting just for the sake of modern use" (ICOMOS China, 2015 (b), p. 103–104).

It is necessary to stress, that both versions of the China Principles (The *Principles for the Conservation of Heritage Sites in China*), namely from 2000 and 2015, invoke authenticity as an essential aspect of the significance of heritage (Zhou, 2018, p. 100). However, the 2015 China Principles particularly stresses the significance of intangible heritage, traditional crafts and the importance of conserving a continuation of cultural traditions by which the conservation of the heritage is integrated into its existing cultural background. Only this approach guarantees the proper maintenance of authenticity, the interpretation considers three aspects: original, traditional craft, and cultural continuity, with an emphasis on cultural diversity (Zhou, 2018, p. 100). Thus, the correct approach covers heritage maintenance, documentation, interpretation of cultural contents and values, continuation of functions and so on. It shows, how the Venice Charter, centered international guidelines has been adapted in China, and reflects the modern Chinese conservation philosophy of caring for both the tangible and intangible, and thus being some kind of a conservation-of-culture theory of authenticity.

Having said that, it is worth noting that there are many parallels between concepts of the living heritage and intangible heritage. After *The Convention for the Safeguarding of the Intangible Cultural Heritage* was adopted by UNESCO in 2003, the theme of 'living heritage' was incorporated soon into preservation activities devoted to intangible heritage. Actually the focus on living heritage, was acknowledged as an essential part of intangible heritage. However, as pointed by Wijesuriya, the conservation approach that can be considered as the living heritage approach is inherently against compartmentalization between tangible and intangible and movable and immovable, and is 'people-centered' where the beneficiaries are both the heritage and the community. Therefore, it may be treated as some kind of combination of the two predominantly existing approaches in contemporary conservation, namely fabric based and values based approaches (Wijesuriya, 2018, p. 44-45).

For living heritage, one the most essential elements is community. Which itself guarantees a historical continuity, manifested in material and spiritual aspects of heritage. As it is underlined in Intangible Heritage Convention, intangible heritage "is transmitted from generation to generation, is constantly recreated by communities and groups in response to their environment, their interaction with nature and their history, and provides them with a sense of identity and continuity" (Wijesuriya, 2018, p. 47). All heritage sites thus, have been continued to survive and have changed. If the continuity is broken, the heritage site, especially historic architectural structures tend to be abandoned by the people. That divorce process from present society results in a particular isolation from the authentic context. If for any reason some value is recognized in the structure, it is kept in its physical form removing it from the continuity of its original use. This can be called the "museumification" (Wijesuriya, 2018, p. 49) of a historic structure, which in consequence impairs the context of authenticity of the site. This phenomenon can be observed above all in Western countries, where community attachment and continuity being the base for living heritage is often disregarded in contemporary society. It is on the contrary, if the historic structure has been adapted over time and meets needs of the present society, and it continues for a particular community to perform functions (use) which are similar (or the same) for which it was created, the aspect of authenticity adds direct value to its nature of living heritage. It needs to be stressed, that all communities undergo constant change. These changes are manifested on the heritage site. Therefore, the conservation of the monument becomes the management of the change. However, to the change the site is also invariably linked continuity. For that reason, it can be said that the process of the protection of heritage is "about the management of continuity and change" (Wijesuriya, 2018, p. 44). Following it, the notion of authenticity could be regarded as the "management of change and continuity". As the historic structure, particularly wooden buildings, manifest the continuity and change of community that is using it, this change is permanently included in its nature. It was pointed already, that continuity is the basis on which to characterize living heritage. This continuity comes from the community continuing a particular function of the building. So this concept is somehow in contrary with modern conservation movements, which have promoted the assumption that it is the duty of the heritage professionals to find a suitable use for heritage under consideration, so best practice is to "use of them for some socially useful purpose". It can be argued that original function has been sometimes replaced by new functions in some heritage buildings. But as long it was done with the intent of the community using the building it seems to be fulfilling the natural process of continuity and thus recognizing authenticity in the dynamics of community continuance – its "living" dimension.

This particular issue is especially applicable to the context of today China, where the rapid development has brought noticeable changes in social lives, cutting out the original communities from the built heritage. The process of "muzeumification" is widely noticeable around the country, where many historic wooden structures have received the status of "monument taken out of history". The problematic assumption of authenticity as maintenance of fabric and intangible values lacks an important dimension of this value – the living community necessary for keeping the site in the constant change resulting directly from the process of continuity. However, as it was stressed, we must not forget it is China where some of the most outstanding examples of living heritage and thus, "living building tradition" can still be observed (Figure 7.9 – 7.27). It is also true regarding rural and urban landscapes considered as living heritage, as an example of the Chinese attempts to preserve national heritage are noticeable on a larger scale. Unfortunately, in many of cases, traditional communities have been prevented access to their heritage as a result modern heritage management systems. Changing areas into "tourist attraction" sites. All of this reflects the growing tensions between the decisions made concerning the preservation of historic wooden architecture in China, and the loss of the dimension of living heritage resulting in a significant loss in the value of authenticity.

Figure 7.9. The "living heritage" still present in a remote Tongle village (near Yezhi), Yunnan Province (photo T. Tomaszek, 2019).

Figure 7.10. The “living heritage” still present in a remote Tongle village (near Yezhi), Yunnan Province (photo T. Tomaszek, 2019).

Figure 7.11. The “living heritage” still present in a remote Tongle village (near Yezhi), Yunnan Province (photo T. Tomaszek, 2019).

Figure 7.12. The "living heritage" still present in a remote Tongle village (near Yezhi), Yunnan Province (photo T. Tomaszek, 2019).

Figure 7.13. The "living heritage" still present in a remote Tongle village (near Yezhi), Yunnan Province (photo T. Tomaszek, 2019).

Figure 7.14. The "living heritage" still present in the Da Luoshui Village on the Lugu lakeshore, Yunnan Province. The village is home to the Mosuo ethnic minority group famous for matriarchal society and their wooden (log) architecture (photo T. Tomaszek, 2019).

Figure 7.15. The "living heritage" of Tibet – prayers in the front of The Jokhang Temple, known as the Qoikang Monastery, a Buddhist temple in Barkhor Square in Lhasa, Tibet Autonomous Region. This temple is considered to be the most sacred and important temple in Tibet. Originally built in 652, and over the next 900 years the temple was enlarged several times with the last renovation completed in 1610 by the Fifth Dalai Lama. Together with the Japanese Hōryū-ji temple is recognized as the world's oldest timber frame Buddhist temple (photo T. Tomaszek, 2018).

Figure 7.16. The "living heritage" of Tibet – prayers in the The Jokhang Temple, known as the Qoikang Monastery, a Buddhist temple in Barkhor Square in Lhasa, Tibet Autonomous Region (photo T. Tomaszek, 2018).

Figure 7.17. The "living heritage" of Tibet – pilgrims around the Main Assembly Hall of Tashi Lhunpo Monastery, Shigatse, Tibet Autonomous Region. The monastery was founded in 1447 by the first Dalai Lama (photo T. Tomaszek, 2018).

Figure 7.18. The "living heritage" of Buddhism in The Tibetan Plateau – The Maitreya Buddha Hall (built in 1379) at Kumbum Monastery (Ta`er Si Temple), Xining, Qinghai Province. It is one of the six major monasteries of the Gelug School of Tibetan Buddhism visited by hundreds of pilgrims every day (photo T. Tomaszek, 2018).

Figure 7.19. The "living heritage" of The Tibetan Plateau – the interior of one of the shops in the Village of Darchen at the foot of Mt. Kailash, Ngari Province, Western Tibet, Purang County, Tibet Autonomous Region. This remote village is a small settlement with mostly nomads passing through the area (photo T. Tomaszek, 2018).

Figure 7.20. Typical wooden country dwellings and a characteristic wooden tower gate – an example of disappearing traditional lifestyle and thus the "living heritage" in Southern China, Mahao Fishing Village, Shizhong District of Leshan, Sichuan Province (photo T. Tomaszek, 2018).

Figure 7.21. Zhaoxing Dong Village – an example of "living heritage" of the ethnic minority of Dong, Guizhou Province. The village was built in the 5th year of the Southern Song Dynasty in 1160A.D. and is famous for wooden drum towers and traditional rice fields. It is considered to be one the most beautiful ancient villages and towns in China (photo T. Tomaszek, 2018).

Figure 7.22. Zhaoxing Dong Village – an example of "living heritage" of the ethnic minority of Dong, Guizhou Province. The village was built in the 5th year of the Southern Song Dynasty in 1160A.D. and is famous for wooden drum towers and traditional rice fields. It is considered to be one the most beautiful ancient villages and towns in China (photo T. Tomaszek, 2018).

Figure 7.23. Zhaoxing Dong Village – an example "living heritage" of the ethnic minority of Dong, Guizhou Province. The village was built in the 5th year of the Southern Song Dynasty in 1160A. D. and is famous for wooden drum towers and traditional rice fields. It is considered to be one the most beautiful ancient villages and towns in China (photo T. Tomaszek, 2018).

Figure 7.24. Zhaoxing Dong Village – an example of "living heritage" of the ethnic minority of Dong, Guizhou Province. The village was built in the 5th year of the Southern Song Dynasty in 1160A.D. and is famous for wooden drum towers and traditional rice fields. It is considered to be one the most beautiful ancient villages and towns in China (photo T. Tomaszek, 2018).

Figure 7.25. Zhaoxing Dong Village – an example of "living heritage" of the ethnic minority of Dong, Guizhou Province. The village was built in the 5th year of the Southern Song Dynasty in 1160A.D. and is famous for wooden drum towers and traditional rice fields. It is considered to be one the most beautiful ancient villages and towns in China (photo T. Tomaszek, 2018).

Figure 7.26. Zhaoxing Dong Village – an example of "living heritage" of the ethnic minority of Dong, Guizhou Province. The village was built in the 5th year of the Southern Song Dynasty in 1160A.D. and is famous for wooden drum towers and traditional rice fields. It is considered to be one the most beautiful ancient villages and towns in China (photo T. Tomaszek, 2018).

Figure 7.27. Zhaoxing Dong Village – an example of "living heritage" of the ethnic minority of Dong, Guizhou Province. The village was built in the 5th year of the Southern Song Dynasty in 1160A.D. and is famous for wooden drum towers and traditional rice fields. It is considered to be one the most beautiful ancient villages and towns in China (photo T. Tomaszek, 2018).

Wooden temples and Sacred Mountains – Chinese tradition and spirituality as a voice in the discourse of authenticity

ABSTRACT: China has a wealth of important heritage sites, many of which are considered sacred or of extraordinary religious importance. Ordinarily, these places host the materialization of religious and spiritual values which takes the form of a temple. These predominately wooden structures, which by the virtue of its properties, as well as intangible symbolic meaning, not only embody the immaterial essence of continuation, but also the cyclical process of destruction and degradation followed by the resulting renewal. Among the holy sites a special place is occupied by the Sacred Mountains. For centuries, these Sacred Mountains have been a site of major significance for the religious traditions of China. At the same time, they were the epicenter of the interpenetration of spiritual values and thoughts, as well as the center of coexistence of man and nature, that for Buddhists or Daoists in itself is a shrine. Wooden architectural structures directly responding to religious and spiritual needs has shaped both the external landscape, closely harmonizing with the natural landscape, but also the internal landscape of the faithful, in which they strengthen the eternal rituals and cyclical power of nature. This chapter, presents a selection of wooden temples from the famous Sacred Mountains in China, while the issue of authenticity in regards to these structures is discussed. This section is depicted by the rich photographic material, displaying the outstanding beauty and spiritual value of the sites.

China has a wealth of important sites many of which are considered sacred or of extraordinary religious importance. Ordinarily, these places are hosts the materialization of religious and spiritual values which takes the form of a temple. These predominately wooden structures, which by the virtue of its properties, as well as intangible symbolic meaning, not only embodies the immaterial essence of continuation, but also the cyclical process of destruction and degradation followed by the resulting renewal. Among the sites of particular importance in China a special place is occupied by the aforementioned Sacred Mountains. Divided into several groups, and are dependant on cosmological or religious determinants. The most renowned five mountains in Chinese history are referred as the Five Great Mountains. They are associated with the supreme God of Heaven and the five main cosmic deities of Chinese traditional religion. The Five Great Mountains depict the five cardinal directions of Chinese geomancy, which includes the center as a direction (Ma, 1998). The mountains have theological and cosmological significance, and thought to represent the ordered world emanating from the God of Heaven (Tian–Shangdi) (Little; Eichman, 2000, p. 148). These five mountains are: Tài Shān (East Great Mountain) in Shāndōng Province, Huà Shān (West Great Mountain) in Shaanxi Province, Héng Shān (South Great Mountain) in Húnán Province, Héng Shān (North Great Mountain) in Shānxī Province and Sōng Shān (Center Great Mountain) in Hénán Province (Robson, 2009, p. 37).

Some of the Sacred Mountains in mainland China and Tibet Autonomous Region hold a special spiritual meaning for Daoists and Buddhists. Both Daoism and Buddhism with their particular philosophies ingrained in Chinese culture, art and daily practice and are recognized as the core of Chinese culture. Daoism is the indigenous religious complex of China and its tradition stretches back thousands of years, as evidenced by the classic texts, the *Daodejing* (*Tao-te Ching*) that was written around 350 B.C. (Pregadio, 2008, p. 1075).

Whereas, Buddhism came to China from India in the first century. Both focus on the spiritual development of the mind and body. In its approach to reality, Daoism considers nature as a model for a balanced life and the pursuit of harmony in the universe as made by constant rhythmic complimentary and opposite forces of yin and yang (Little; Eichman, 2000, p. 148). Buddhism emphasizes the interconnectedness of the past, present and future. Apart from some similarities and differences between them, both are still the backbone of Chinese civilization and the cyclical concept of time.

It is important to understand that throughout China's history, both Buddhist as well as Daoist pilgrims have gone to mountains seeking spiritual sustenance and solace, thus, the mountains have become a most holy destination. They believed that is was the mountain peaks in particular that were closest to heaven and the gods, and the ideal training ground in the pursuit of enlightenment and transcendence. Some mountains were considered sacred per se, as they were considered the incarnation of the Buddha or were used as burial sites. Some other peaks were also worshiped by various religions and have become a most important pilgrimage destination, as is the case with, for example, Mount Kailash in Tibet. Therefore, with time, certain mountains in China became closely associated with Daoist and Buddhist pilgrimage (Wang, 2016, p. 173). While there are numerous mountains throughout China that are considered sacred, nine of them in mainland China achieved particular prominence. Among these nine sacred mountains five are preeminent for Daoists and four are paramount to Buddhists. The five sacred Daoist mountains are Tai Shan, in Shandong province; Hua Shan, in Shaanxi province; Heng Shan Bei, in Shaanxi province; Heng Shan Nan, in Hunan province; and Son Shan, in Henan province. The four sacred Buddhist mountains are Emei Shan, in Sichaun province; Wutai Shan, in Shaanxi province; Jiuhua Shan, in Anhui province; and Putuo Shan, in Zhejiang province. In the period of its greatest prosperity, the sacred mountains in China supported hundreds of monasteries and sheltered elaborate temples, cliff inscriptions and stone tablets. They were also associated with major works of Chinese poetry and art. Unfortunately, mostly due to the historical and political circumstances of 20th Chinese history – particularly the 1949 Communist Revolution, which established an atheist state and then The Cultural Revolution of the 1960s, which continued state-sanctioned suppression of archaic religious ideas – approximately 90 percent of the temples were destroyed and the sacred mountains underwent a difficult time of trying to keep spiritual activities alive (Little; Eichman, 2000, p. 148).

The holiest mountain for the Daoists is Tai Shan, where it was established the first known Daoist temple in China in 351 B.C. Today, there are only 22 temples remaining from the hundreds, that at some point in history, were protecting the slopes of the mountain (Little; Eichman, 2000, p. 148). For Buddhism, the holiest of sites are Emei Shan along with Wutai Shan. Emei Shan is home to the first Buddhist temple in China erected in the first century and also of the world's largest Buddha statue. Unfortunately, only 20 monasteries survive from once more then 100 standing on the slopes in Emei Shan (Figure 8.1 – 8.26). Wutai Shan is the highest mountain with five peaks in northern China and it is considered the actual place where Buddhism was brought directly from India and thus it has been an active sacred spot for pilgrims for 2,000 years. There are fifty-three monasteries remaining on Mount Wutai, including the East Main Hall of Foguang Temple which is the highest ranking timber building to survive from the Tang Dynasty, and the Ming Dynasty Shuxiang Temple with a huge complex of 500 'suspension' statues, representing Buddhist stories woven into three dimensional pictures of mountains and water. The temple buildings on Mount Wutai present a catalogue of the way Buddhist architecture developed and influenced palace building over a wide part of China (Figure 8.27 – 8.50). Mount Wutai remains an important repository of Buddhist culture, and attracts a variety of pilgrims from all across Asia, who apart from spiritual reasons, also visit for the extensive library of books collected by Emperors and scholars. The temples are inseparable from their mountain landscape – their high peaks snow covered for much of the year, thick forests of vertical pines, firs, poplar and willow trees and lush grassland (Mount Wutai (China), UNESCO).

Figure 8.1. Shanmen, Baoguo Temple ("Nation Dedication Temple") established in the Wanli period (1573–1619) of the Ming dynasty; Éméi shān (Mount Emei), one of the Four Sacred Buddhist Mountains of China (traditionally regarded as the bodhimaṇḍa, or place of enlightenment, of the bodhisattva Samantabhadra), Sichuan Province (photo T. Tomaszek, 2018).

Figure 8.2. Mahavira Hall, Baoguo Temple ("Nation Dedication Temple") established in the Wanli period (1573–1619) of the Ming dynasty; Éméi shān (Mount Emei), one of the Four Sacred Buddhist Mountains of China, Sichuan Province (photo T. Tomaszek, 2018).

Figure 8.3. The Main Hall at Fuhu Temple (or Tiger Taming Temple) it is the largest nunnery in Mount Emei originally built in the Tang Dynasty (618-907), destroyed at the beginning of the Qing dynasty (c. 1644), in 1651 a group of disciples started to rebuild the temple over approximately, the next twenty years; Éméi shān (Mount Emei), one of the Four Sacred Buddhist Mountains of China, Sichuan Province (photo T. Tomaszek, 2018).

Figure 8.4. Dormitories for monks at Fuhu Temple (or Tiger Taming Temple) is the largest nunnery in Mount Emei originally built in the Tang Dynasty (618-907), destroyed at the beginning of the Qing dynasty (c. 1644), in 1651 a group of disciples started to rebuild the temple over approximately, the next twenty years; Éméi shān (Mount Emei), one of the Four Sacred Buddhist Mountains of China, Sichuan Province (photo T. Tomaszek, 2018).

Figure 8.5. Dormitories for monks at Fuhu Temple (or Tiger Taming Temple) is the largest nunnery in Mount Emei originally built in the Tang Dynasty (618-907), destroyed at the beginning of the Qing dynasty (c. 1644), in 1651 a group of disciples started to rebuild the temple over approximately, the next twenty years; Éméi shān (Mount Emei), one of the Four Sacred Buddhist Mountains of China, Sichuan Province (photo T. Tomaszek, 2018).

Figure 8.6. A statue of Arya Avalokiteshvara surrounded by 500 Buddhist Arhats, the Arhat Hall, the Fuhu Temple (or Tiger Taming Temple) is the largest nunnery in Mount Emei originally built in the Tang Dynasty (618-907), destroyed at the beginning of the Qing dynasty (c. 1644), in 1651 a group of disciples started to rebuild the temple over approximately, the next twenty years; Éméi shān (Mount Emei), one of the Four Sacred Buddhist Mountains of China, Sichuan Province (photo T. Tomaszek, 2018).

Figure 8.7. Hongchunping Temple, originally built in the Song Dynasty (960–1279) and extended in the late Ming Dynasty (1368–1644); Éméi shān (Mount Emei), one of the Four Sacred Buddhist Mountains of China, Sichuan Province (photo T. Tomaszek, 2018).

Figure 8.8. The Main Hall, Hongchunping Temple, originally built in the Song Dynasty (960–1279) and extended in the late Ming Dynasty (1368–1644); Éméi shān (Mount Emei), one of the Four Sacred Buddhist Mountains of China, Sichuan Province (photo T. Tomaszek, 2018).

Figure 8.9. Dormitories for monks, also provides accommodation for pilgrims to this remote area (mountains and dense forest), Hongchunping Temple, originally built in the Song Dynasty (960–1279) and extended in the late Ming Dynasty (1368–1644); Éméi shān (Mount Emei), one of the Four Sacred Buddhist Mountains of China, Sichuan Province (photo T. Tomaszek, 2018).

Figure 8.10. Located among dense forest on the hills of Mount Emei the Leiyin Temple, originally dating to the Ming dynasty, the former Moksha temple was renamed Leiyin in 1884; Éméi shān (Mount Emei), one of the Four Sacred Buddhist Mountains of China, Sichuan Province (photo T. Tomaszek, 2018).

Figure 8.11. Leiyin Temple, originally dating to the Ming dynasty, the former Moksha temple was renamed Leiyin in 1884; Éméi shān (Mount Emei), one of the Four Sacred Buddhist Mountains of China, Sichuan Province (photo T. Tomaszek, 2018).

Figure 8.12. Leiyin Temple, originally dating to the Ming dynasty, the former Moksha temple was renamed Leiyin in 1884; Éméi shān (Mount Emei), one of the Four Sacred Buddhist Mountains of China, Sichuan Province (photo T. Tomaszek, 2018).

Figure 8.13. Shenshui Pavilion, formerly the provincial governor Wu Yongxian's villa and renamed Shengshui Temple in the Wanli period (1573-1620) of the Ming Dynasty, then renamed Shenshui Pavilion in the Shunzhi period (1643-1661) of the Qing Dynasty; Éméi shān (Mount Emei), one of the Four Sacred Buddhist Mountains of China, Sichuan Province (photo T. Tomaszek, 2018).

Figure 8.14. Shenshui Pavilion, formerly the provincial governor Wu Yongxian's villa and renamed Shengshui Temple in the Wanli period (1573-1620) of the Ming Dynasty, then renamed Shenshui Pavilion in the Shunzhi period (1643-1661) of the Qing Dynasty; Éméi shān (Mount Emei), one of the Four Sacred Buddhist Mountains of China, Sichuan Province (photo T. Tomaszek, 2018).

Figure 8.15. Shenshui Pavilion, formerly the provincial governor Wu Yongxian's villa and renamed Shengshui Temple in the Wanli period (1573-1620) of the Ming Dynasty, then renamed Shenshui Pavilion in the Shunzhi period (1643-1661) of the Qing Dynasty; Éméi shān (Mount Emei), one of the Four Sacred Buddhist Mountains of China, Sichuan Province (photo T. Tomaszek, 2018).

Figure 8.16. Shenshui Pavilion, formerly the provincial governor Wu Yongxian's villa and renamed Shengshui Temple in the Wanli period (1573-1620) of the Ming Dynasty, then renamed Shenshui Pavilion in the Shunzhi period (1643-1661) of the Qing Dynasty; Éméi shān (Mount Emei), one of the Four Sacred Buddhist Mountains of China, Sichuan Province (photo T. Tomaszek, 2018).

Figure 8.17. Shenshui Pavilion, formerly the provincial governor Wu Yongxian's villa and renamed Shengshui Temple in the Wanli period (1573-1620) of the Ming Dynasty, then renamed Shenshui Pavilion in the Shunzhi period (1643-1661) of the Qing Dynasty; Éméi shān (Mount Emei), one of the Four Sacred Buddhist Mountains of China, Sichuan Province (photo T. Tomaszek, 2018).

Figure 8.18. The interior of one of the halls, Shenshui Pavilion, formerly the provincial governor Wu Yongxian's villa and renamed Shengshui Temple in the Wanli period (1573-1620) of the Ming Dynasty, then renamed Shenshui Pavilion in the Shunzhi period (1643-1661) of the Qing Dynasty; Éméi shān (Mount Emei), one of the Four Sacred Buddhist Mountains of China, Sichuan Province (photo T. Tomaszek, 2018).

Figure 8.19. The roof construction of one of the halls, Shenshui Pavilion, formerly the provincial governor Wu Yongxian's villa and renamed Shengshui Temple in the Wanli period (1573-1620) of the Ming Dynasty, then renamed Shenshui Pavilion in the Shunzhi period (1643-1661) of the Qing Dynasty; Éméi shān (Mount Emei), one of the Four Sacred Buddhist Mountains of China, Sichuan Province (photo T. Tomaszek, 2018).

Figure 8.20. Xianfeng Temple, once called the Ciyan Temple, first built in 1281. The temple sits in front of Huayan Hill and it is backed by rugged cliffs and surrounded by a fantastic landscape; Éméi shān (Mount Emei), one of the Four Sacred Buddhist Mountains of China, Sichuan Province (photo T. Tomaszek, 2018).

Figure 8.21. Xianfeng Temple, once called the Ciyan Temple, first built in 1281. The temple sits in front of Huayan Hill and it is backed by rugged cliffs and surrounded by a fantastic landscape; Éméi shān (Mount Emei), one of the Four Sacred Buddhist Mountains of China, Sichuan Province (photo T. Tomaszek, 2018).

Figure 8.22. Xianfeng Temple, once called the Ciyan Temple, firstly built in 1281. The temple sits in front of Huayan Hill and it is backed by rugged cliffs and surrounded by a fantastic landscape; Éméi shān (Mount Emei), one of the Four Sacred Buddhist Mountains of China, Sichuan Province (photo T. Tomaszek, 2018).

Figure 8.23. Xixiangchi Temple, literally means "Elephants` Washing Pool", also known as Tianhua Chanyuan, one of the most important Buddhist temples on Mount Emei. It was founded in the Ming dynasty and was expanded during the reign of the Kangxi Emperor in the Qing dynasty. Éméi shān (Mount Emei), one of the Four Sacred Buddhist Mountains of China, Sichuan Province (photo T. Tomaszek, 2018).

Figure 8.24. The Entrance Gate, Xixiangchi Temple, literally means "Elephants` Washing Pool", also known as Tianhua Chanyuan, one of the most important Buddhist temples on Mount Emei. It was founded in the Ming dynasty and was expanded during the reign of the Kangxi Emperor in the Qing dynasty. Éméi shān (Mount Emei), one of the Four Sacred Buddhist Mountains of China, Sichuan Province (photo T. Tomaszek, 2018).

Figure 8.25. Xixiangchi Temple, literally means "Elephants' Washing Pool", also known as Tianhua Chanyuan, one of the most important Buddhist temples on Mount Emei. It was founded in the Ming dynasty and was expanded during the reign of the Kangxi Emperor in the Qing dynasty. Éméi shān (Mount Emei), one of the Four Sacred Buddhist Mountains of China, Sichuan Province (photo T. Tomaszek, 2018).

Figure 8.26. Xixiangchi Temple, literally means "Elephants' Washing Pool", also known as Tianhua Chanyuan, one of the most important Buddhist temples on Mount Emei. It was founded in the Ming dynasty and was expanded during the reign of the Kangxi Emperor in the Qing dynasty. Éméi shān (Mount Emei), one of the Four Sacred Buddhist Mountains of China, Sichuan Province (photo T. Tomaszek, 2018).

Figure 8.27. The wooden archway, The Dailuoding Temple (built in the Ming Dynasty, repaired and reconstructed in the following Ming and Qing Dynasties), Wutai Shan (Mount Wutai, also known as Mount Qingliang), one of the Four Sacred Buddhist Mountains in China, Shanxi Province (photo T. Tomaszek, 2019).

Figure 8.28. Hall of Sakyamuni, The Dailuoding Temple (built in the Ming Dynasty, repaired and reconstructed in the following Ming and Qing Dynasties), Wutai Shan (Mount Wutai, also known as Mount Qingliang), one of the Four Sacred Buddhist Mountains in China, Shanxi Province (photo T. Tomaszek, 2019).

Figure 8.29. Monks praying in the Hall of Sakyamuni, The Dailuoding Temple (built in the Ming Dynasty, repaired and reconstructed in the following Ming and Qing Dynasties), Wutai Shan (Mount Wutai, also known as Mount Qingliang), one of the Four Sacred Buddhist Mountains in China, Shanxi Province (photo T. Tomaszek, 2019).

Figure 8.30. Manjusri Hall, Luohou Temple, first established in the Tang dynasty (618-907) and initially called Shanzhu Geyuan. It was rebuilt in 1492 during the Hongzhi period of the Ming dynasty, then restored during the Wanli period (1572-1620) and in 1705 converted to Tibetan Buddhism. In 1792, it was partially reconstructed; Wutai Shan (Mount Wutai, also known as Mount Qingliang), one of the Four Sacred Buddhist Mountains in China, Shanxi Province (photo T. Tomaszek, 2019).

Figure 8.31. The Hall of Sakyamuni, Luohou Temple, first established in the Tang dynasty (618-907) and initially called Shanzhu Geyuan. It was rebuilt in 1492 during the Hongzhi period of the Ming dynasty, then restored during the Wanli period (1572-1620) and in 1705 converted to Tibetan Buddhism. In 1792, it was partially reconstructed; Wutai Shan (Mount Wutai, also known as Mount Qingliang), one of the Four Sacred Buddhist Mountains in China, Shanxi Province (photo T. Tomaszek, 2019).

Figure 8.32. The Interior of the Hall of Sakyamuni, Luohou Temple, first established in the Tang dynasty (618-907) and initially called Shanzhu Geyuan. It was rebuilt in 1492 during the Hongzhi period of the Ming dynasty, then restored during the Wanli period (1572-1620) and in 1705 converted to Tibetan Buddhism. In 1792, it was partially reconstructed; Wutai Shan (Mount Wutai, also known as Mount Qingliang), one of the Four Sacred Buddhist Mountains in China, Shanxi Province (photo T. Tomaszek, 2019).

Figure 8.33. The Gate Tower entrance, Jile Temple area, Nanshan Temple complex, originally built in the 13th century, then expanded during the 16th and 19th centuries; Wutai Shan (Mount Wutai, also known as Mount Qingliang), one of the Four Sacred Buddhist Mountains in China, Shanxi Province (photo T. Tomaszek, 2019).

Figure 8.34. The Hall of Sakyamuni, Jile Temple area, Nanshan Temple complex, originally built in the 13th century, then expanded during the 16th and 19th centuries; Wutai Shan (Mount Wutai, also known as Mount Qingliang), one of the Four Sacred Buddhist Mountains in China, Shanxi Province (photo T. Tomaszek, 2019).

Figure 8.35. The Hall of Thunder, Youguo Temple area, Nanshan Temple complex, originally built in the 13th century, then expanded during the 16th and 19th centuries; Wutai Shan (Mount Wutai, also known as Mount Qingliang), one of the Four Sacred Buddhist Mountains in China, Shanxi Province (photo T. Tomaszek, 2019).

Figure 8.36. The Hall of Four Heavenly Kings, Puhua Temple, without verification of when it was originally established, is also called "The Temple of Jade Emperor". The present temple was completed in 1925 during the Republic of China period (1912-1949); Wutai Shan (Mount Wutai, also known as Mount Qingliang), one of the Four Sacred Buddhist Mountains in China, Shanxi Province (photo T. Tomaszek, 2019).

Figure 8.37. The Hall of Sakyamuni, Puhua Temple, without verification of when it was originally established, is also called "The Temple of Jade Emperor". The present temple was completed in 1925 during the Republic of China period (1912-1949); Wutai Shan (Mount Wutai, also known as Mount Qingliang), one of the Four Sacred Buddhist Mountains in China, Shanxi Province (photo T. Tomaszek, 2019).

Figure 8.38. The Hall of Three Buddhas, Puhua Temple, without verification of when it was originally established, is also called "The Temple of Jade Emperor". The present temple was completed in 1925 during the Republic of China period (1912-1949); Wutai Shan (Mount Wutai, also known as Mount Qingliang), one of the Four Sacred Buddhist Mountains in China, Shanxi Province (photo T. Tomaszek, 2019).

Figure 8.39. The interior of the Hall of Four Heavenly Kings, Puhua Temple, without verification of when it was originally established, is also called "The Temple of Jade Emperor". The present temple was completed in 1925 during the Republic of China period (1912-1949); Wutai Shan (Mount Wutai, also known as Mount Qingliang), one of the Four Sacred Buddhist Mountains in China, Shanxi Province (photo T. Tomaszek, 2019).

Figure 8.40. The Hall of Sakyamuni, the Pusading Temple, first established in the reign of Emperor Xiaowen of Northern Wei (467–499) and initially called "The Great Temple of Manjushri", in 631 (the Tang dynasty) it was rebuilt by Buddhist monk Fayun and renamed "The Zhenrong Temple", then in 1581 renovated by the government; Wutai Shan (Mount Wutai, also known as Mount Qingliang), one of the Four Sacred Buddhist Mountains in China, Shanxi Province (photo T. Tomaszek, 2019).

Figure 8.41. The Hall of Sakyamuni, the Pusading Temple, first established in the reign of Emperor Xiaowen of Northern Wei (467–499) and initially called "The Great Temple of Manjushri", in 631 (the Tang dynasty) it was rebuilt by Buddhist monk Fayun and renamed "Zhenrong Temple", then in 1581 renovated by the government; Wutai Shan (Mount Wutai, also known as Mount Qingliang), one of the Four Sacred Buddhist Mountains in China, Shanxi Province (photo T. Tomaszek, 2019).

Figure 8.42. The Hall of Sakyamuni, the Pusading Temple, first established in the reign of Emperor Xiaowen of Northern Wei (467–499) and initially called "The Great Temple of Manjushri", in 631 (the Tang dynasty) it was rebuilt by Buddhist monk Fayun and renamed "Zhenrong Temple", then in 1581 renovated by the government; Wutai Shan (Mount Wutai, also known as Mount Qingliang), one of the Four Sacred Buddhist Mountains in China, Shanxi Province (photo T. Tomaszek, 2019).

Figure 8.43. The Hall of Manjusri, the Shuxiang Temple, built in early years of of the East Jin Dynasty, rebuilt in the Tang, Yuan (in the year 1325), and Ming Dynasties (in the year 1496); Wutai Shan (Mount Wutai, also known as Mount Qingliang), one of the Four Sacred Buddhist Mountains in China, Shanxi Province (photo T. Tomaszek, 2019).

Figure 8.44. The Tayuan Temple, one of the five famous temples for practicing Chan Buddhism in China, originally a stupa of Xiantong Temple, named the Great White Pagoda, was constructed in 1302, then in 1407 it was expanded to a temple and got its name; Wutai Shan (Mount Wutai, also known as Mount Qingliang), one of the Four Sacred Buddhist Mountains in China, Shanxi Province (photo T. Tomaszek, 2019).

Figure 8.45. The Zushi Hall, the Tayuan Temple, one of the five famous temples for practicing Chan Buddhism in China, originally a stupa of Xiantong Temple, named the Great White Pagoda, was constructed in 1302, then in 1407 it was expanded to a temple and got its name; Wutai Shan (Mount Wutai, also known as Mount Qingliang), one of the Four Sacred Buddhist Mountains in China, Shanxi Province (photo T. Tomaszek, 2019).

Figure 8.46. Pilgrims circling the Great White Pagoda, the Tayuan Temple, one of the five famous temples for practicing Chan Buddhism in China; Wutai Shan (Mount Wutai, also known as Mount Qingliang), one of the Four Sacred Buddhist Mountains in China, Shanxi Province (photo T. Tomaszek, 2019).

Figure 8.47. The Tayuan Temple, one of the five famous temples for practicing Chan Buddhism in China; Wutai Shan (Mount Wutai, also known as Mount Qingliang), one of the Four Sacred Buddhist Mountains in China, Shanxi Province (photo T. Tomaszek, 2019).

Figure 8.48. The Manjusri Hall, the Wanfoge Temple; Wutai Shan (Mount Wutai, also known as Mount Qingliang), one of the Four Sacred Buddhist Mountains in China, Shanxi Province (photo T. Tomaszek, 2019).

Figure 8.49. The Manjusri Hall, the Wanfoge Temple; Wutai Shan (Mount Wutai, also known as Mount Qingliang), one of the Four Sacred Buddhist Mountains in China, Shanxi Province (photo T. Tomaszek, 2019).

Figure 8.50. The Hall of Avalokitesvara – interior, the Xiantong Temple, first established in the Yongping period (58–75) in the Eastern Han dynasty (25–220) and initially called "Dafu Lingjiu Temple", expanded and renamed as "The Huayuan Temple" in the period of the Northern Wei dynasty (386–534), then reconstructed in the period of Emperor Taizong (599–649) of the Tang dynasty (618–907) with the name of "Great Huayan Temple", next renovated and renamed "Great Xiantong Temple" in the period of Hongwu (1368–1398) in the early Ming dynasty (1368–1644), restored in the 1950's and 1970's in the period of the People's Republic of China; Wutai Shan (Mount Wutai, also known as Mount Qingliang), one of the Four Sacred Buddhist Mountains in China, Shanxi Province (photo T. Tomaszek, 2019).

Modern China is a country that has undergone an enormous social and economic transition. Together with a new found consciousness towards national heritage and thus national identity. A new movement has emerged postulating to rebuild the temples destroyed during the 20th century and to develop a tourism strategy that plainly banks on the religious history of China's sacred mountains. This brings a number of new problems, as the development of tourism poses a serious threat to the proper protection of the remains of the Sacred Mountains.

In recent decades, the state government has made various attempts towards protecting this important heritage. As a result Tai Shan, Emei Shan and Wutai Shan were inscribed on the UNESCO World Heritage list. However, all the conservation efforts are in tension with powerful profit motives. For example, in 2003, the Chinese government pulled down hundreds of illegal structures on Tai Shan, including food huts, souvenir stands and billboards. At the same time, unfortunately, the Chinese government supports the creation of theme parks on the mountains seeing them as beneficial for developing tourism and increasing the economic attractiveness of the area. On other hand, in 1995, the ARC (Alliance of Religions and Conservation) began working with the China Daoist Association, and later with the Chinese Buddhist Association, to develop conservation programs aligned with and supported by religious belief and practice (Noga, 2008). Together they have helped create management programs for the Daoist mountains Hua Shan and Tai Shan, and assist conservation efforts for the Buddhist Wutai Shan and Emei Shan.

Among Sacred Mountains in Tibet Autonomous Region, probably the most worshipped is Mount Kailash (Tibetan: གངས་རིན་པོ་ཆེ). It is the second highest peak of the Gangdise Mountains, located in Burang County, south of Linma Pangyong Co and Laang Co. At its foot, is the northern source of the Shiquan River and in the upper reaches the source of the Indus River (Allen, 1982, p. 67). Tibetan Buddhism believes that this mountain is the navel of the universe and the residence of Shengle King, representing immeasurable happiness, and it is also the place where Padmasambhava defeated the Bons. Mount Kailash is known as Mount Meru in Buddhist texts. It is central to its cosmology, and a major pilgrimage site for some Buddhist traditions (Buswell, 2004, p. 407-408). Whereas, followers of Jainism believe that the mountain is the place where its ancestor Lisapa Nath Ṛṣabha-nātha gained the Tao. Hinduism believes that the mountain is Shiva's residence and the center of the world. According to legend, Bon religion also originated there. Therefore, for the purpose of pious belief and praying for blessings and fighting bad deeds, many believers of the above-mentioned religions have worked hard to make pilgrimages around the mountains from all directions. As a result, Mount Kailash sees an endless stream of pilgrims and it has become a veritable "multiple" religious shrine. It is said, that after one will circle Mount Kailash 10 times they can then avoid the suffering of hell in the next 500 reincarnations, and after 100 times can ascend to heaven and become a Buddha in this life (Camaria, 1996). Buddhists and Hindus usually go around the mountain in a clockwise direction, while Jain and Bon believers go around the mountain in a counterclockwise direction. Due to its religious significance, Kailash remains an unclimbed mountain. There are still remaining some temples in the footsteps of Kailash, which are used for pilgrims and play essential role in the cultural landscape of this sacred site (Figure 8.51 – 8.55).

For centuries, the sacred mountains have been a site of major significance for the religious traditions of China. At the same time, they were the epicenter of the interpenetration of spiritual values and thoughts, as well as the center of coexistence of man and nature that for Buddhists or Daoists in itself is a shrine. Wooden architectural structures directly responding to religious and spiritual needs has shaped both the external landscape, closely harmonizing with the natural landscape, but also the internal landscape of the faithful, in which they strengthen the eternal rituals and cyclical power of nature.

It would not be an exaggeration to say that cultural heritage is the soul of the nation. Therefore, heritage could be understood and experienced somehow as the soul of the nation (Zerrudo, 2018, p. 108). Heritage, according to Zerrudo, may also be understood as

Figure 8.51. The Direpu temple (Dirapuk monastery) located on the foothills of Mt. Kailash, Tibet (photo T. Tomaszek, 2018).

Figure 8.52. Mt. Kailash seen from The Direpu temple (Dirapuk monastery), Tibet (photo T. Tomaszek, 2018).

Figure 8.53. The Direpu temple (Dirapuk monastery) located on the foothills of, Tibet (photo T. Tomaszek, 2018).

Figure 8.54. The Direpu temple (Dirapuk monastery) located on the foothills of, Tibet (photo T. Tomaszek, 2018).

Figure 8.55. Pilgrims at The Direpu temple (Dirapuk monastery) located on the foothills of, Tibet (photo T. Tomaszek, 2018).

anything of value from the past that gives particular meaning and identity to the present and at the same time would inspire generations to come (Zerrudo, 2008). Wooden architecture creates a special type of heritage, in case of which, the fabric used for the materialization of ideas and beliefs stay strictly connected to the circular nature of reality. This is particularly true in the case of China, where wooden building is the most important manifestation of local building traditions and spiritual beliefs.

In the spatial arrangement of Buddhist wooden temples in China, that reflect the particular cosmology of Chinese spirituality, an important role is played by religious theory. The temples can be understood in Chinese tradition, similarly like in numerous South Asian-originating religions, as a sacred cosmological mountain, the *maha meru* (Mount Meru – the dwelling of the Gods) (Peou, 2018, p. 203), which is the axis of the world. Therefore all the architectural structures, including also secondary ones such as stupas, lanterns, and *budo* (stupas for monks) are located in a specified order to express various areas with respect to the Buddhist pantheon as described in various stupas (Kim, 1999, p. 63–66). It must be remembered, that this is an essential aspect of aesthetic value, as such a relationship between Buddhist beliefs and the materialized layout of a temple is the most essential factor of authentic form (Lee, 2018, p. 61). Without this understanding, it is not possible to consider what is the real authentic form in a Chinese context. Therefore, the form of a building should not be frozen in a particular time in the search for the authenticity, as its notion, the genuine continuity of a building can be guaranteed only when there is a transmission of sublime thought, spirit, and philosophy from one generation to another through its constant materialization. Considering the relationship between the tangible and intangible elements of Chinese wooden temple buildings, it is not possible to fully transmit values by pursuing only the earliest form (Lee, 2018, p. 62). So continuity also includes change, and change as a natural process of continuity includes nature transmission.

Due to the spiritual, as well as historic significance of sacred places with wooden heritage in China, for example the Sacred Mountains with the wooden temples or Buddhist wooden temples in Tibet Autonomous Region, there are numerous approaches to view, experience and also interpret these sites. Each of possible perception resulting from these approaches have the same validity and importance for believers and must be incorporated into the overall understanding of this place. The main challenge is to cater to all the requirements of sacred use of the space, including the use of the space by pilgrims coming from diverse cultures. At the same time, properly safeguarding the material part of the built heritage. All is becoming highly challenging, as the pilgrims must be free to show their devotion through various activities which may also directly impact the heritage structures. It must be stressed, that a conflict of interest arises when the heritage building, namely wooden temple or sacred object itself, becomes an object of veneration. For example, certain pilgrims may break off parts of the temple fabric to show devotion to the deity. Others, may give offerings such as food, perfume, milk, gold leaf, and coins, or burn incense on or next to the scared structure. When discussing the authenticity of a site and wooden structures like this, there are numerous conflicts that arise between is ritual use and the intended preservation of material object. This issue determines the history of a belief system that still persists, that for example of Buddhism in mainland China or Tibet (Weise, 2018, p. 128). Therefore, it must be underlined the importance of the focus on the continuation of tradition and intangible aspect, and not solely on the preservation of material remains. The attributes of the value of this places, thus its authenticity, are on the one hand, the material objects of actual temples devoted for particular rites. As well as on the other hand, the site's significance as a place of pilgrimage and as an overall sacred spot. Significant sacred places worthy of mention are Mount Kailash together with its Kora pilgrimage and Sacred Mountains for instance Wutai Shan or Emei Shan. Therefore, pilgrimage in its purest form is the essence of these sacred spots, where materialization of the wooden temple acts only as a formalization of the spiritual belief frozen in ephemeral body of architectural structure.

However, it must also be remembered, that these sacred sites go through the process of constant change and adaptation consistent to the concept of cyclical nature of time. Thus, the phenomenon of taking form of regular and constant adaptation, has itself provided the possible thread of continuity required for the survival of these historic structures and intangible spiritual aspect of the sites. It becomes very problematic then to reconcile the necessary renewal with an approach which is acceptable within the understanding of authenticity, therefore to pursue appropriate renewal involving conflicting challenges and expectations. What becomes even more challenging, as without any break in the function of the temple that is a living monument, it is almost impossible to determine its specific state needed to determine the scope of possible preservation.

It is very likely, that the tradition of conservation of sacred sites in China, effectively forming a specific mix of preservation, reconstruction and restoration, began along with the construction of the earliest religious images and the architectural structures that were meant to house them. With passing of time, these buildings were developed and changed multiple times. This has allowed them to continue over centuries until this day. It is understood the aims of the conservation appears to have two causes, firstly assuring the continuity of spiritual and religious activity and secondly to guarantee the longevity of the artifacts or edifices used in this religious activity (Tiwari, 2009, p. 5). Therefore conservation, being the cycles of construction, destruction, renovation and/or reconstruction has led to overlapping various layers of history, meanings and materials, forming what we call wooden heritage of sacred sites in China. However, it also established its own standards of understanding, knowing and safeguarding it for presentation, preservation and enhancement, thereby informing us how and what action were taken over time to conserve the holy sites and temples passed on by the preceding ones (Chapagain; Tiwari, 2018, p. 137). These various approaches for conservation, which appeared over time, were really a consequence of the architectural forms that were developed in the building traditions and meanings given to the space of the temple. Therefore, the very architecture that shapes over time is based on climatic and religious circumstances in that and no other way, consequently, has resulted in such and no other shaping of the approach for conservation, forming the present image of holy places, such as Wutai Shan. So the conservation of these places can be named somehow as the culture of reconstruction and adaptation for religious continuity.

From the point of view of the spiritual values of the Buddhist understanding of the essence of the material sphere of the temple, the exclusive purpose of the form is associated directly somehow with the protection which it provides for the enshrined within the temple body a religious object or sacred relic (De Vos, 2018, p. 165). As the form of the wooden temple could evolve and thus change in some details over time (even if it were maintaining some general design solution of Chinese temple architecture), the form should not to be considered critical. Greater importance is attached to the continuation of the original authenticity in the enshrined sacred object/relic (thus religious function), so that the temple credibility was maintained and protected by assuring continuity of the architectural body, not necessarily the continuity of the original fabric or even shape. Any additions introduced later in the history of the temple were never treated as disturbing the credibility of the original sacred space function that established authenticity; rather they reinforced the legitimate value attached to the original enshrinement (De Vos, 2018, p. 167). Therefore, in this understanding, the temple does not lose its authenticity when it is constantly restored or rebuilt. What is important, each rebuild was conducted as in the first building phase, with similar expressions of religious fervor. However, it needs to be noted that in Asian tradition, as is the case of Japanese culture, the temple was rebuilt using timber extracted from sacred forests, where it was specially grown for the purpose of temple maintenance, thus in this way, respecting and protecting the authenticity as an intransient quality and ensuring the essential heritage requirements for maintaining the essence of religious values. This approach, based on constant rebuilding maintains the visual value expressed by Buddhist spirituality, where completeness of the temple is necessary for proper spiritual feeling. It could be said that the longevity of temples was somehow related to the life of the sun and moon, as long as they last, which means forever, thus as a wooden temple should be considered more of a sacred site than an architectural structure constituting his material body (Department of Archaeology, 1904, p. 167). The very disturbance and removal of the spiritual function from a temple irreversibly destroys the symbolic value attached to the enshrinement. On this basis, the temple building deprived its sacred meaning and function has little value and has lost its original authenticity and therefore, could be even permanently removed from the point of Buddhist spirituality and cosmology.

Sacred wooden architecture, represents an inseparable link between the spiritual and the material form. This true statement is particularly true in regards to China. Wooden temples express complicated patterns of spiritual and religious complexes, embodied in the broader spectrum of cultural forms and processes that show the dynamic interrelationship between tangible and intangible. Spirit or tradition cannot thus be separated from form, design, layout, material and craftsmanship as it is underpinned– it is from the spiritual belief that the temple or sacred object emerges. Without spiritual value, the material value may have limited meaning. (Sharma, 2018, p. 178).

An essential part of sacred wooden structures is the painted and sculpture decoration which is inseparably connected to the architectural form. Therefore all must be treated as an essential unity of its spiritual and cultural message. It should be stressed, that in Buddhist tradition, sacred paintings such as temple decoration, are seen as tools for realization and thus have a mandate reason for their creation. Additionally, the source of painting itself is given divine origin, stated in the early sacred Indian manuscripts that later influenced Buddhism in China. This essential aspect once again established the inseparable connection between the spiritual and the material (Sharma, 2018, p. 174).

Some of the ancient texts mentioned above were at some point translated from Sanskrit into Tibetan, and while the original Indian manuscripts were already lost, they are the testimony of this tradition. These Tibetan translations form part of so-called the Tangyur volumes, which are the *Citralaksana* and *Pratimāmānalakshanām*. The first deals with painting while the second with image-making. The theory of proportions followed in image-making and painting discussed in both the texts *Pratimāmānalakshanām* and *Citralaksana* enable an artist, who is trained to it, to be able to both create new paintings as well as renew damaged ones. What is particularly interesting, an artist (with knowledge of the these texts) can recreate a damaged or partially lost image based only on for example a single finger and its setting in a larger composition. As noted by Sharma, in this essential knowledge lies continuity and the process by which it is transferred into creating sacred art (Sharma, 2018, p. 174). So there is a knowledge of proportion in books, but it is still overtaken by the fact that the image is a creation/the way of giving life by Divine, having a Divine essence (because the first image is like that). Not as to paint is substantial, but the knowledge of how to use it Divine, how to pass the Divine. The proportions are constant, which is important as it shows the circularity included in it. This raises the problem of the copy and its understanding in the tradition of Western thought. In this case, it is difficult to call something that is repeated and remade wholly as a copy.

If the painting was not executed in the proper way the deity could have refused to reside in it at the time of consecration. Therefore, the senior monks conducted the consecration and if the work was not considered to be of standard the painting most likely was to be rejected. The "intent" of the creator or artist is what imparts the specific intangible quality to the painting (Sharma, 2018, p. 176). To reduce poor quality painting, traditional texts refer to and describe what is translated as "blemishes in painting" which can cause misfortune or loss to befall on the village or householder. Similarly, hallmarks of excellence are noted too (Sharma, 2018, p. 174). Finally, it needs to be added that both, the creator and the viewer, perceive the painted image according to their own state of inner realization. The spiritual attributes inform the attributes of material and substance, form and design, use and function and without the former, the value of the latter ceases. As noted by Sharma, therefore assessing and articulating conditions of authenticity of attributes of spirit and feeling become all the more critical (Sharma, 2018, p. 174), particularly from the point of outside observer.

The essence of wood, continuity, change and preservation as a religious practice

ABSTRACT: It can be said without any doubt, that wooden architecture symbolizes the spiritual and material well-being of a civilization. This statement applies particularly to the Chinese civilization, where wooden structures, particularly wooden temples, are an exceptionally important materialization of local tradition achieved in vernacular architecture. The need for the continuous repairs to the wooden shrines has resulted in the continuity of hundreds of years of traditional living patterns and development of the socio-economic system bound up with the temple. Within these traditions, particular notions of the issue of authenticity persist, through ideas such as the process of renovation which is somehow deeply connected to religious beliefs of rebirth and a circular nature of time. As per the Buddhist philosophy, the repair work done for the temple gives merit for both the present and future lives. To fulfill the religious needs, the structures of the temple or a monastery complex should therefore be maintained in a way of complete renewal and necessitates permanent beautification. This chapter discusses preservation practices with a deep connection to religious beliefs in the manifestation of change and continuity, as well as the intangible essence of wood.

According to Schaeffer, in Buddhist tradition (particularly on Tibetan Plateau) artists or architects engaged in the creation of sacred art and architecture must have knowledge that significantly exceeds that is ordinarily expected for their field. Their study, and thus further knowledge base, had to extend to the five major classical forms; the fine arts (*bzo rigpa*), inner knowledge (*nang rigpa*), logic (*gtantshigs rigpa*), language (*sgra rigpa*) and medicine (*gsoba rigpa*). Additionally, they also needed to study the five minor forms: poetry (*snyan ngag*), prosody (*sdeb sbyor*), synonyms (*mngon brjod*), drama (*zlos gar*) and astrology and divination (*skar tsis*). Tibetans took over this model of the five forms of knowledge from the Sanskrit treatises on Indian Buddhism as early as the ninth century (Schaeffer, 2011, p. 294-296).

There is a long and rather rich history of texts and traditions relating to the creation and the creators of sacred art and architecture in the Indian subcontinent, that further influenced Buddhism in China. Apart from the aforementioned *Pratimāmānalakshanām* of sage Atreya on image-making translated from Sanskrit into Tibetan between the seventh and eleventh centuries (Jugnu, 2013), another worthy of reference is the *Mayamata* composed between the early ninth to late twelfth century, which devotes an entire chapter to renovation works (Dagens, 1995). In the passage 35.40b-42 it states: "When a temple (...) or images are to be renovated, the work is always to be done with materials similar (to those used initially) or better ones and never with less good ones. In the case of a decrepit (object), the knowledgeable one wishing to restore it, will process as indicated above in order to return it to its original condition according to rule; (but), if the object was small it is desirable that it be restored to at least equal size or even to a larger one for that is always auspicious". Further, in the passage 35.43 it is stated: "That which is made smaller has to be made of better materials or may be made identical, with the same materials as previously and with its dimensions in accordance with those calculated from the sanctum, the pillars and doors of the sanctum" (Dagens, 1995, p. 339). Later the renovation of images is clarified: "(...) an image which is incomplete is to be rejected immediately and a new image installed in its place. An image of the required height and thickness, but which is split or which has any flaws of that kind, must be rejected and another image installed in its place according to rule. A metal or earth image lacking hands, nose, adornment, ears or teeth is to be restored to its original condition but if it is a principal limb which is missing it must be thrown out and a new one put in its place" (Dagens, 1995, p. 338–339).

As mentioned before the *Pratimāmānalakshanām* similarly discusses the treatment of damaged images, and suggests that the wood to "besmeared with ghee or honey and placed in fire". The text concludes with the caring of old and damaged images prescribing ritual destruction and reinstating of the sacred essence in a new image (Jugnu, 2013). This kind of belief of permitting badly damaged sacred images to decay naturally is still observed in Buddhist tradition in China, particularly in Tibet. It is well worth noting here that both preservation as well as decay are accepted in the life cycle of a sacred object (Jugnu, 2013, p. 176). Citing Gelong Tsewang Rigzen, "repair and renovation of the monastery is an ongoing process since we live and pray in them" (Dogra, 2008, p. 12). This expresses Buddhist understanding of the circularity of time and therefore the essence of continuity. However, as change is inherently included in continuity, the decay and passing of matter is a natural part of the entire cycle.

In order to gain insight into the depths of a Chinese cultural and spiritual context, it is necessary to challenge the notion of destruction, and thus decay, as is understood in Western tradition based mainly on rationality, and revisit it from an Eastern perspective, as deeply connected to the everyday existence of people through their beliefs in Buddhism and Hinduism. Within these traditions, particular notions of the issue of authenticity persist, through ideas such as the process of renovation which is somehow deeply connected to religious beliefs of rebirth (Dhumal, 2018, p. 184) and circular nature of time.

This issue of destruction and decay applies also to decorative painting of wooden temples. It is then usual that only the sections of the decorative painting which were partially lost are created anew, while the parts of the decoration survive in good condition are not initially touched (Sharma, 2018, p. 177). Therefore, for some outside observers the substantial difference in patina of old painting and brightness of the fresh colors used in the new paintings could be seriously distracting. However, from the point of spirituality and Buddhist tradition, this re-touched painting is still considered to be original, and the differences coming from old a new paint are without any disturbance of authenticity of the temple (Figure 6.1 – 6.26). It is also interesting, that there is no distinction between the old and new painting in terms of their primary value as depictions of the Buddha worthy of worship. As noted by Sharma, preserving Buddhist images, whatever their age, is viewed as an act of merit. Citing the late Rinpoche Dzongsar Ngari Chödje, "The older a monastery, the better is the accuracy of iconography of statues and paintings, and most essentially, the greater the spiritual value because they were blessed and consecrated by those great old masters who possessed immense energy which they infused into these places, creating reserves of energy in these places which one could still feel" (Sharma, 2018, p. 177-178).

The forementioned conservation practice, particularly due to the frequently encountered, constant reconstruction over such long periods, means that almost none of the wooden heritage structures can be regarded as "original" as a whole or even in part, if we narrow the definition of "original" to the initial construction and initial finalized shape. Thus no one of these structures fulfil the test for authenticity as the form frozen in some "original" version. Therefore, it is very hard to make any specific historic reference that would form the sole basis of judging authenticity as majority of the temples show multiple historic, architectural or symbolic layers, thus making it quite difficult to infer the originality both in terms of material as well as design. As noted by Tiwari, "for cultures, whose historical development is predominantly steered by religions, that had and have a cyclical and seasonal notion of time, history becomes diachronic and the layers seasonal. When such accumulated seasonal layers span centuries (...) conservation becomes a practitioner's nightmare particularly if we take the definition of authenticity at layers of the seasonally renewed originals" (Tiwari, 2009, p. 9–10).

There is one more issue that makes all of this even more complicated, and reference to authenticity becomes extremely challenging. There are some temples, where a special part of the architectural structure or particular inner space is restricted from any form of outsider access or even its documentation (in some cases even verbal descriptions about what is there is prohibited). In a such situation, where traditional religious and cultural practices prevent an outsider access, especially a conservation expert, from knowing or inspecting the material, construction, design or any other aspect of a sacred space, there are no clear reference from which authenticity can be formulated. Thus, it needs to be underlined, that in the case of these religious structures, authenticity lies solely with what the initiated and authorized priest sees, believes, knows and eventually does to the structure (Chapagain; Tiwari, 2018, p. 135). As long as we deal with living religious heritage, this issue will have a particular meaning in China, especially due to Buddhism capturing the meaning and use of religious spaces.

The search for authenticity encounters difficulties which show little usefulness of its application to the tradition of wooden temples in sacred sites in China. Perhaps, authenticity should be treated as relevant to continuous repair, reconstruction and even beautification, but following certain standards of material and craftsmanship available only for authorized priests or religious individuals. Therefore, it gives some sense of authenticity through design and construction skill practiced as trade and the experience passed on from ritual and religious predecessor or respected spiritual tradition. Thus, perhaps for justifying the authenticity, would be to recognize that this very tradition, which by its very nature, prevents us from thinking about authenticity just through particular aspects including material, design etc. is an essential aspect of authenticity in this particular case and that no other aspects can be outweighed (Chapagain; Tiwari, 2018, p. 136).

It is worth mentioning, essential connections of Buddhism to the tree and its spiritual meaning as a living body and as an axis mundi – axis of Earth. It is said that Siddhartha Gautama, the spiritual teacher who became known as the Buddha, have attained enlightenment or *Bodhi* circa 500 BCE under a tree, the Tree of Awakening (Bodhi Tree or Bodhi Fig Tree) located in Bodh Gaya, Bihar, India (Gethin, 1998, p. 22).

In the following centuries, the spiritual message of Buddhism was transferred to other parts of the World, it often was brought keeping connotations to the sacred tree. It happened for example in Sri Lanka, where Buddhism was introduced by Arhat Mahinda (son of Emperor Asoka) from India in the second century BCE, who together with her sister Arhat Sangamitta arrived with a sapling from the sacred Bodhi tree and with artisans excelled in 18 different arts and crafts to serve the Bo tree shrine (Geiger, 1912, p. 37). Therefore, the sacred tree could be considered as the embodiment of Buddha, thus the most important carrier of a spiritual message and a wooden temple at the same time as the enshrining body of sacred message and thus sacred site (Figure 9.1).

It seems that this tradition gained unparalleled importance in Asia, as sacred trees have gained a particular status in the religious pantheon. They continue to hold profound meaning in the Far East, particularly in China, where they are believed to be inhabited by spirits or deities and where they are thus worshiped. They are viewed as being literal sanctuaries and they are protected as a physical and spiritual embodiment of the divine nature. Technically, any tree can become a sacred tree through a ritual process of inviting a spirit to inhabit it. However, usually those considered sacred reveal special properties, such as age or appearance. Apart from individual sacred trees, shrines and Buddhist temples are quite often surrounded by sacred forests (often called "sacred groves"), which are believed to be the dwelling places of ancestral spirits (Senda, 1992, p. 131). So this connection to the living tree provides a special meaning when it is used as a fabric for the erection of a temple. Firstly, it is a fabric that comes from a once (and somehow still) alive body, thus going through a continuous cycle of successive birth and death, secondly, it is deeply connected to the

Figure 9.1. Buddha head overgrown by a fig tree; Wat Maha That, Ayutthaya, Thailand (photo T. Tomaszek, 2019).

alignment and spiritual message of Lord Buddha. Therefore, the use of wood as a fabric for the shrine may be understood as not accidental. As mentioned above, artisans excelled in 18 different arts and crafts to serve the Bo tree shrine, who arrived in Sri Lanka with Arhat during introduction of Buddhism there, and who seem to have counterpart artisans working traditionally on wooden Baddish temples in China, confirm this thought. By their knowledge and skills, passed from generation to generation, it was possible to keep the continuity of circular nature of Buddhist reality handed over in the body of a wooden shrine. This may partly explain the importance of the wooden temple for Buddhism and give an idea as to why wood became such an important material for erecting shrines. At the same time, the very physical form of the temple is of no importance until the meaning of the spiritual message is placed in it, that is, until it becomes a home for the deity. Here, too, lies the dual nature of wood and its direct connection with nature's driving cycles.

It can be said therefore, that Buddhist wooden built heritage is a combination of both the tangible and intangible aspects – interrelated and interdependent. The traditions of the societies keeping continuity alive, has led to the survival of monuments and in other ways the wooden shrines have caused a continuation of spiritual patterns and meanings of beliefs. In this light, conservation practices of the temples also need to be viewed, those that have been known in China for millennia. As per the Buddhist philosophy, the repair work done for the temple gives merit for both the present and future lives (Kalupahana, 1975). It seems to be an essential aspect for understanding the continuity of the temple structures actualizing the living nature of spiritual message.

It needs to be underlined, that one of the most essential aspects to fulfil the Buddhist religious needs, is a clear idea about the importance of maintaining (thus conserving and restoring) the structures of the temple or a monastery complex in a way of complete renewal and the necessary permanent beautification. Therefore, it is in contrary to international standards prioritizing the minimal intervention and quite often maintenance the historic value of the "archeological" state of the heritage building. Of course, in the case of a significant part of the wooden temples in China in general the minimum intervention method is used for conservation, or in the other words, method of conservation without completion, and these projects are usually totally funded by the government (Kulatunga, 1999). This kind of maintenance (or conservation), is done in accordance to international standards influencing Chinese practice, but not really following the religious needs and traditional concept of circular nature of time. What is more, in several cases these religious structures, conserved with minimal intervention, generally need to be further maintained after restoration by the government, as they did not fulfil the Buddhist religious needs of the believers (Ratnyake, 2018, p. 150). For the religious reasons the devotees wish to contribute to the repair work of the total completion of the temple, as it gives a certain spiritual feeling to the devotees – clear and clean feeling in the mind. Buddhists believe that this feeling gives merit to a better life especially in the next birth. At the same time, everyday maintenance (like removal of overgrowing plants and other types of activities) on an incomplete temple, does not provide such feelings, as such shrine does not have the same visual impact. In general, as noted by Perera, Buddhist devotees believe that those completed shrines maintain the religious appearance, thus are more positive places for worship. What even more important, in the disciplinary texts for Buddhist monks, the Lord Buddha stated that it is not suitable for monks to live in dilapidated buildings (Perera, 2016).

The need for the continuous repairs to the wooden shrines has resulted in continuity for the hundreds of years of traditional living patterns and development of the socio-economic system bound up with the temple. Thus, the life of communities belonging to the temple was connected heavily to the daily rituals and annual festivals, which additionally guaranteed up keep and maintenance of the temples. However, due to rapid changes of lifestyle in China, an introduction of an open economy for instance, development of communication and transportation, better education, or even new international relations, long standing religious and cultural patterns are changing very quickly and becoming disintegrated from daily routine. This particularly applies to Baddish religious communities attached to wooden temples. Where there has been a huge shift in lifestyle, resulting in wider career options. Meaning that many people from traditional societies have parted ways with family professions and converted to pursue other vocations. This has resulted in the growing lack of skills necessary for traditional repair work on temples and keeping them in "spiritual shape" according to religious needs. As a consequence of this, the required skills for the tangible aspects are also disappearing from the whole community. Resulting in a major collapse in the self-sufficient traditional conservation system. It has a direct impact on challenge of maintaining the authenticity of wooden Buddhist temples in China, particularly in the areas such as Yunnan province or Tibet Autonomous Region. Due the introduction of legal ownership, authority and protection by the state, combined with changing socio-economic values, many of the traditions are vanishing and continuity of authenticity in traditional conservation of wooden temples in China facing enormous challenge. Any attempts to continue the existence of these shrines stripped of traditional beliefs and necessary skills for maintenance threatens to condemn them to museum-like architectural objects. Therefore, it is necessary to turn towards the community and the traditional systems, other than solely practicing modern technical methods which may be alien to long surviving cultural and religious systems. And what is also important, the wooden shrines cannot be restrained from living culture activities, such as these practiced by the believers and pilgrims. They must have a full access to conduct all the religious activities, such as praying or giving offerings. The former particularly has its significance, as Buddhists respect relics similar to a living Buddha, and thus visitors to the temples with relics must have the possibility to deliver their offerings, even if they could potentially impair the historic fabric of heritage structure. It needs to be underlined again, that this way the living traditional conservation practice will be continued, as people participate in these works with great enthusiasm, in the belief of collecting merit (Ratnyake, 2018, p. 151).

It can be said without any doubts, that wooden architecture symbolizes the spiritual and material well-being of the civilization. This statement applies particularly to Chinese civilization, where wooden structure is an exceptionally important materialization of local tradition achieved in vernacular architecture. Among wooden structures, as the most significant wooden temples stand out, being a direct expression of the most important intangible assets. The protection of this heritage is therefore vital for continuity of the culture and spiritual tradition. It needs to be remembered, that majority of wooden temples in China are living religious heritage ("living Buddhist temples"), thus, attempts to raise certain issues with regard to the application of the concept of authenticity must be understood in the context of living religious heritage. It is crucial to raise the appropriateness of placing an equal emphasis to all aspects related to authenticity for sustaining the spiritual values of a living religious site. To discuss authenticity, all the appropriate attributes of the wooden temple must be recognized. Wherein the "attributes" should be understood as aspects of which are associated with or express values holistically. In the other words, all attempts must be made to identify the attributes related to a particular wooden temple, as they are vital to the understanding of authenticity, and in consequence protection, conservation and management of the site.

According to Weerasinghe, there are three essential physical or tangible domains of Buddhist heritage, particularly wooden temples, as aspects that are related to authenticity: space, form and sight (vision). Recognizing these aspects is significant for any attempt at conservation of Buddhist heritage. Weerasinghe is noting therefore, that the space of Buddhist heritage, a temple in a broader sense, living heritage sites must be able to accommodate Buddhist rituals; the form of such living heritage must be able to embody the liturgical and ritual values; and the sight should be able to visually convey the meaning and values particularly significant for this religious site (Weerasinghe, 2006, p. 183). Accordingly, it could be stated, that a high degree of authenticity is expressed by use and function and spirit and feeling (Cooray, 2012, p. 112). As that most of the international conservation documents, following the principles of the Venice Charter, have adopted a mostly so-called archaeological approach to heritage monuments, and therefore do not adequately address the concepts of a living heritage site and its authenticity, this particularly reflects difficulties with applying such an understanding in a Chinese context. Therefore, any approach to the protection of Buddhist heritage, particularly in the case of Chinese wooden temples, should focus on maintaining a balance between archaeological and spiritual characteristics of the site, so that they serve and continue their religious function. To properly recognize these aspects is essential in any attempts to identify the issue of authenticity of wooden built heritage in China. Even though the original physical form and design of the majority of temples has been altered over several periods in history. It was carried out through restoration, renovation and rehabilitation. This has helped facilitate a natural continuity and thus evolving religious practices as well as the practical needs of succeeding periods. These interventions have not affected the authenticity in use and function as well as the sprit and feeling, therefore, cannot be treated as constituting a falsification of the authenticity of the architectural structure. In the majority of cases, the most prominent Buddhists or clergy demanded that the wooden temples be restored or completely reconstructed in order to enhance the spiritual values and to re-establish the Buddhist identity (Cooray, 2018, p. 157). It must be also remembered, that the array of monasteries consisting of wooden structures which facilitated different functions ranging from meditation to scholarly activities of monks, showcase an unparalleled development and evolution over time in the whole of the Buddhist world, thus also in China. Therefore, in the case of living heritage, these wooden temples, carry high degrees of spiritual values, seems to be essential context for the discussion on authenticity. Throughout history, the form of the temples has evolved with the introduction of new tangible as well as intangible items. Such as new forms of rituals, for this reason the authenticity of this aspect is not to preserve the same items forever, which will lead it to a static situation, but to introduce from time to time, new items that are compatible with its overall historic character (Cooray, 2018, p. 161). However, now a new challenge is being faced, increased tourism. How will it be possible to maintain a balance between growing religious pilgrimage and tourism, so that controlled tourism is not to have any effect on the sanctity of the sacred spaces of these heritage wooden temples.

Balancing authenticity and interpretation – contemporary Chinese practice in the protection of wooden architectural heritage

ABSTRACT: Due to massive economic development and the growing need for maintaining and recognizing national identity, the Chinese approach to the protection of heritage changed drastically at the beginning of the 21st century. The meaning of "the old" has undergone a significant modification, and from the understanding of it as term of implying something as backward and hidebound, it has changed to an appreciation of its own tradition, culture, heritage and national identity. This understanding of "the old" however, does not always comply with local practices in present day China. Practice shows quite a different notion of this term and thus diversity in its understanding and applications of authenticity. All over China one can see re-creations of "the old" wooden buildings and large scale reconstructions of architectural complexes, manifested by "antique" products, particularly executed as "archaized buildings" or "ancient villages", which imitate historic buildings. The most likely need for this practice, arises from people's nostalgia for their heritage, which appeared in the course of rapid modernization and industrialization. This chapter discusses practices in protection and conservation of wooden architectural heritage noticeable in modern China. With a particular focus on the issue of reconstruction and recreation, thus re-inventing traditions, as well as the elusive border between the original and the copy.

To gain a better understanding of the modern Chinese approach to cultural heritage and the notion of authenticity, it is necessary to look back at what has been happening historically in the country during the past 70 years. The turning point was the creation of the People's Republic of China in 1949. From that time, the official government agenda for monument preservation has focused on recognizing "the achievement of the ancient working people in architectural creation and to critically evaluate the national tradition and embrace its good aspects, thereby developing a new socialist structure" by "constructing the new and demolishing the old" (Liang, 2001, p. 35). As a consequence, during the period of 1966–1976, called as the Cultural Revolution, the majority of monuments such as temples or palaces were regarded as the unnecessary relics of a feudal system of organized superstition. As it was a time of building the new order in China, most of the remaining old system monuments were demolished. This approach changed drastically in the last decade of 20th century. The Chinese government then implemented a new open-door policy and the country started its continuous economic reform. All of this combined has brought quickly growing public sentiment towards national culture and history, this has resulted in an urgent call for the revitalization of the country's cultural identity. As a response to this new situation, a huge number of heritage (re) construction projects at historical sites has been conducted throughout China (Zhu, 2017, p. 192). It marked the beginning of an appearance of a new trend, the search for economic profits. This has focused on developing tourism and a real estate industry at the sites of cultural and historical significance. This has led to the rise of new investment companies, as well as new activities of local governments and agencies, directed towards heritage development for local recreational and commercial use and with it economic benefits. As noted by Agnew, all together it affected and diversified the criteria for authenticity (Agnew, et al., 2004, p. 41). Referring to Zhu, local understandings of authenticity are not the results of a passive adoption of global ideas, but rather the conglomerate of global and national concepts in dynamic negotiation with aspects of homogenization and heterogenization (Zhu, 2015, p. 603). Therefore, it can be said that we can observe different forms of heritage practice that have emerged in China. As well as the diversity of authenticity applications, showing a range of localized understandings of its notion, postulated by different players in the network of heritage conservation. Also various approaches to its interpretation as a part of protection of the built heritage, understood as the platform for maintaining and recognizing national identity.

The ICOMOS Ename Charter ("Charter for the Interpretation and Presentation of Cultural Heritage Sites") in 2008 raised important matter in the discussion on the significance of interpretation as part of the protection of cultural heritage and the struggle to maintain authenticity in correlation with interpretation (ICOMOS, 2017). In the preamble of this document, it is identified that "interpretation and presentation are essential components of heritage conservation efforts and as a means of enhancing public appreciation and understanding of cultural heritage sites". The charter specifies that, in principle, interpretation must come from verifiable sources and respect the authenticity of each heritage site, thus it need to be applied without distorting their [the monument's] meanings (Poshyanandana, 2018, p. 81). Therefore, talking about authenticity brings us to the combining of various aspects, where the apparent contradiction of authenticity of materials or authenticity of architectural form stands contra authenticity of setting, forming one holistic notion of this value. Thus, it is a necessary consideration on the overall integrity of form, only as suffices for comprehension and aesthetic balance, where symbolic religious, social or spiritual meanings have become comprehensible and interpretable and the landscape features of the past are maintained or properly restored.

According to Majid, the conservation practice in Asia largely focuses on the subject rather than the object, thus the intangible often supersedes the tangible (Majid, 2018, p. 51). The concept of authenticity may be regarded as having therefore a non-homogeneous notion, at the same time involving awareness on the contemporary role of historic areas, as most monuments went through changes at different periods and occasions. But because changes to these monuments were constructed a long time ago to fulfil their changing spiritual needs, we have accepted them as authentic. The above addresses the problem, that the obsession with form results in a building that is frozen in time (Lee, 2018, p. 56). Furthermore, it remains still problematic, whether authenticity is a useful concept in the discourse on conserving heritage in an Asian context.

What needs to be noted, apart from difficulties with the translation of the terms like authenticity into Asian languages and culture, particularly Chinese, it is also impossible to obtain clear meaning for various other terms, such as monument, site, setting, protection and conservation. These terms are difficult to accurately translate into an Asian languages, even if they have a clear meaning in the languages of the West, therefore their notion always brings us to numerous connotations and the cultural intricacies (Zhan, 2018, p. 93).

Referring to the Liang Sicheng and his attempts to define the original state or original form, some professionals interpret the original state as the state when a monument was first completed or in its prime time. Resulting in the practice of the so-called authoritative restoration, which promotes the elimination of other historical information that is considered non-original (Zhan, 2018, p. 91). In the 1980s, and heavily based on academic research, the style of architecture from a specific period attracted more attention, while local influence, as well as later modifications were almost completely ignored (Zhou, 2018, p. 97). Thus, it can be said that the original style was considered when it was to a certain extent cleared from historic layers. Additionally, more or less around this time, the 1950s Soviet Union renovation style strongly influenced Chinese conservation practice, focusing mostly towards pursuing the original state or the most glorious image of an historic building (Zhou, 2018, p. 97).

The Chinese introduction of an open-door policy, has resulted among others, in wide ranging involvement with international heritage conservation movements, thus also an adoption of concepts emerging in the international arena. Therefore, step by step, China has also recognized the significance of "the old" and the "value of history". Being influenced by years of a Maoist perspective on cultural heritage, it was required to rethink and redefine this borrowed idea of "the old". Dominant voice of Liang postulate of "restoring the old as it was", took the key position in understanding heritage "authenticity", particularly in the case of historic architecture preservation (Zhu, 2017, p. 193). At the same time, the meaning of "the old" has undergone a significant modification, and from understanding it as term implying something backward and hidebound, changed to appreciation of its own tradition, culture, heritage and national identity. However, this understanding of "the old" does not always comply with local practices in present day China, which practices show quite a different notion of this term. In any part of the country, we can come across some kind creations of "the old", manifested by "antique" products, particularly executed as "archaized buildings", which are imitating historic buildings. These structures are built in masses and reproduced in a different context, to attract tourists and visitors (Figure 10.1 – 10.8). Such architecture, as noted by Zhu, suggest a "past", playing a significant role in a marketing strategy, being created in order to develop local business (Zhu, 2017, p. 193). So often, these buildings are designed for the purpose of representing ancient architecture, with symbols and decor are purposely used to adorn them in a classical way for this building type, are erected in reality only to be attractive for tourists and visitors, and are used mainly as souvenir shops and tourist accommodation.

Figure 10.1. Creation of "the old" manifested by "archaized buildings", which are imitating historic wooden buildings of a particular style – Tianjin's Ancient Culture Street (Guwenhua Jie), a pedestrian pathway complex dotted with temple gates and kiosks on the west bank of the Hai River, Tianjin. It was opened in 1986, as a splendid replica of classic folk architecture it preserves the architectural style of the Qing dynasty (1644-1911). It is particularly designed as a place for visitors to experience Chinese folk custom, and as such, contains examples of nearly all local culture in one place. The perfect example of the re-creation of "the old" in strengthening of national identity (photo T. Tomaszek, 2019).

Figure 10.2. Creation of "the old" manifested by "archaized buildings", which are imitating historic wooden buildings of a particular style – Tianjin`s Ancient Culture Street (Guwenhua Jie), a pedestrian pathway complex dotted with temple gates and kiosks on the west bank of the Hai River, Tianjin (photo T. Tomaszek, 2019).

Figure 10.3. Creation of "the old" manifested by "archaized buildings", which are imitating historic wooden buildings of a particular style – Tianjin`s Ancient Culture Street (Guwenhua Jie), a pedestrian pathway complex dotted with temple gates and kiosks on the west bank of the Hai River, Tianjin (photo T. Tomaszek, 2019).

Figure 10.4. Creation of "the old" manifested by "archaized buildings", which are imitating historic wooden buildings of a particular style – Tianjin`s Ancient Culture Street (Guwenhua Jie), a pedestrian pathway complex dotted with temple gates and kiosks on the west bank of the Hai River, Tianjin (photo T. Tomaszek, 2019).

Figure 10.5. Creation of "the old" manifested by "archaized buildings", which are imitating historic wooden buildings of a particular style – Tianjin`s Ancient Culture Street (Guwenhua Jie), a pedestrian pathway complex dotted with temple gates and kiosks on the west bank of the Hai River, Tianjin (photo T. Tomaszek, 2019).

Figure 10.6. Creation of "the old" manifested by implementing "archaized buildings" among preserved historical structures – Pingyao (Pingyao Ancient City), Shanxi Province. This settlement is famed for its well-preserved Ming and Qing urban planning and architecture, and the streets and storefronts still largely retain their historical appearance. It is also due to organized restoration that has been undertaken periodically since the 15th century, the most recent phase beginning in the past decades (photo T. Tomaszek, 2018).

Figure 10.7. Creation of "the old" manifested by implementing "archaized buildings" among preserved historical structures – Pingyao (Pingyao Ancient City), Shanxi Province (photo T. Tomaszek, 2018).

Figure 10.8. Creation of "the old" manifested by implementing "archaized buildings" among preserved historical structures – Pingyao (Pingyao Ancient City), Shanxi Province (photo T. Tomaszek, 2018).

Bearing in mind the recommendations of the China Principles, this tendency has nothing to do with what is described in the document for heritage conservation and authenticity. The value of authenticity of buildings like these, is not attached to any historical meaning or archeological evidence. Still it cannot be easily denied, that it comes directly as a response to the Chinese people's quest for authenticity and the visitors' imaginations, that authorize this common practice of "restoring the old" in this case represented by recreation. It can be almost indisputably said, that the need for this practice arises from the people's nostalgia for their heritage, which appeared in the course of rapid modernization and industrialization happening in China over the past few decades.

To discuss authenticity, originality is a hard bypassed concept in the context of Chinese heritage conservation. The debate between originality and the present-state reflects the logical conflict between traditional and modern conservation ideas (Zhou, 2018, p. 98). Recently in China, quite different practical conservation tendencies can be observed. These could be viewed towards excessive as well as expensive, interventions to restoration works, mostly in the form of large-scale reconstructions. A perfect example of this kind of approach, could be illustrated in the reconstruction of Datong city, Shanxi province. In most of these cases of reconstructed heritage, it poses the serious threat that the false tends to become mixed with the genuine. (Zhan, 2018, p. 91). Heavily controversial, and close to the peak of the dispute on understanding the concepts of original state and current state in Chinese practice, is a widely observed trend of "reviving the past glory" of a monument. Focused on restoring a given monument to a condition, as close as possible to its imagined ideal state of its former expected splendid appearance. Some regard this trend to be more of an attempt to satisfy certain political or functional needs and could bring disastrous results for the monument in question (Zhan, 2018, p. 91).

The need the reconstructions of historic wooden structures, some wooden compounds or villages might be questioned, particularly since these do not seem to be linked to the living tradition or material culture of any specific communities, but they are rather a contemporary trend of interpreting the perceived golden age of Chinese material culture (Weise, 2018, p. 123). It may however, also be seen as rebuilding a golden age and a cultural identity that has been destroyed or, better said, weakened after the Cultural Revolution. From a theoretical point of view the contemporary reconstructions like these aiming at re-creating the wooden heritage in China to its former most glorious state, somehow provide justification for needs of the test of integrity which demands "wholeness and intactness" (UNESCO, 2016, paragraph 88) of heritage to be considered authentic (Figure 10.9 – 10.13).

Figure 10.9. Restoring "the old" by re-creation of "archaized buildings" (even if they are "fake antiques"). Fulfilling the people's nostalgia for their heritage, Wenshufun, Chengdu, Sichuan Province (photo T. Tomaszek, 2018). The inscription (in English) on the information sign located on the entrance to Wenshufun: "Wenshu District, a commercial courtyard built in the Traditional Folk Houses of West Sichuan Area, is an embodiment of history, culture and commerce. Combined with its themes of Buddhist and folk cultures, it fully presents us with the humanistic and historical essence of old Chengdu. What`s more with its natural beauty and deep cultural load, this combination of modern Commercial Street with the traditional courtyard, completely overturned the conventional form of Commercial Street".

In fact, the various approaches were introduced to fulfil the realities of the reconstruction of wooden heritage in living village or urban spaces, which as complexes in constant change had throughout history gone through the process of "cyclical renewal" (Weise, 2018, p. 127). There are examples of projects carried out based on anastylosis, renovation, integration and conjecture. What needs to be stressed, that a lot of buildings had their size and function changed since their original construction, and in the case of many projects new techniques were introduced for restoring the historic buildings.

The attitude towards reconstruction of architectural heritage in China, is underlined in the recommendations of the Article 22 of the Law of the People's Republic of China on Protection of Cultural Relics (The Law of the People's Republic..., 2007). Here it is stated as follows: "Where immovable cultural relics are totally damaged, the ruins shall be protected

Figure 10.10. Restoring "the old" by re-creation of "archaized buildings" (even if they are "fake antiques"). Fulfilling the people's nostalgia for their heritage, Wenshufun, Chengdu, Sichuan Province (photo T. Tomaszek, 2018).

Figure 10.11. Restoring "the old" by re-creation of "archaized buildings" (even if they are "fake antiques"). Fulfilling the people's nostalgia for their heritage, Chengdu, Sichuan Province (photo T. Tomaszek, 2018).

Figure 10.12. When the false tends to become mixed with the genuine – "a reinvention of tradition" as the people's nostalgia for their heritage and strengthening of national and ethnic identity – A contemporary train station near Zhaoxing Dong Village, Guizhou Province, built in the traditional architectural style of the ethnic minority group of Dong (photo T. Tomaszek, 2018).

Figure 10.13. When the false tends to become mixed with the genuine – "a reinvention of tradition" as the people's nostalgia for their heritage and strengthening of national and ethnic identity – A row of houses on the entrance to the Zhaoxing Dong Village, Guizhou Province, built in the traditional architectural style of the ethnic minority group of Dong (photo T. Tomaszek, 2018).

and the damaged relics may not be rebuilt on the original site." However, there are some exceptions, which are discussed with the State Administration of Cultural Heritage. When the special circumstances arise, the particular relic is to be rebuilt on the original site. It mostly applies to single monuments. In the case of justifying necessity to rebuild a major site that is protected at a national level, the matter is dealt with by the People's Government of the relevant province, or autonomous region, or municipality directly under the Central Government, which must have an approval of their decision from the State Council (Zhu, 2017, p. 194).

The consent for reconstruction of historic monuments was supported by Zhewen Luo, Liang's former assistant and successor of his philosophy on architectural conservation. However, according to Luo, the only proper and justified reconstruction, as well as proper restoration, should be based on "scientific evidence" (Luo, 1999, p. 20). This requirement is now included in official law established by the Chinese Government regarding conservation of monuments. According to this law, the reconstruction of historical buildings and monuments are permitted, only in the case when sufficient scholarly and scientific information on all historical periods and styles that are in question are available.

However, the actual heritage reconstructions that have been carried out in China in the last two decades, show different practices. An illustrative example of the issue is the case of the Old Town of Lijiang, Yunnan Province. The historical center of Lijiang is one of the most famous clusters of traditional architecture of Naxi people, an ethnic group that inhabited this spot during the Yuan Dynasty (1271–1368). In 1997, it was inscribed on the UNESCO World Heritage List. This fact has brought many changes to this site; growing commercialization and outflow of original residents to outside the city center. Soon after, there were projects implemented, aiming at the reconstruction of some of the important architectural structures in the city. One of them was Mu's Residence. Tusi Mu was the ruler of Naxi during the Yuan (1271-1368), Ming (1368-1644), and Qing (1644-1911) dynasties. During their reign, the residence was the center of politics, power and wealth in ancient Naxi. The original buildings of the residence were destroyed together with the majority of surrounding houses during warfare in the Qing Dynasty. The final demolition happened during the Cultural Revolution of the 1960s (UNESCO, undated, Old Town of Lijiang).

The complex of the residence was rebuilt between 1997 and 1999 on the original site of the former buildings financed by the World Bank. The idea was to capture the magnificent image of what life was like at this great palace. As noted by Zhu, Mu's Residence, now one of the main tourist attractions in Lijiang, is not an academically proper reconstruction, but it is more "a reinvention of tradition". As an architectural complex it was reconstructed with no historical information about the original structures. What is more, the historical construction principles of the Naxi culture were not even abided to. However, for the local government this reconstruction is regarded as an exemplary pilot project and is heavily marketed to maximize its local benefit. What is particularly interesting, the local government allegedly drew upon the concept of "restoring the old as it was", nevertheless made very special reinterpretation of the notion of this idea (Zhu, 2017, p. 195).

Observing the contemporary trends in the reconstruction of wooden historic complexes it could be noted, that the main objective seems to be overall preservation of the physical environment and character of the chosen historic settlement. Thus, it is intended that the main architectural character of the historic settlement is preserved and areas of special architectural and historic interest are protected. Unfortunately, not always the intangible aspects are maintained with the same care, such as the preservation of local traditional skills and keeping alive traditional cultural activities. Therefore, in the majority of cases the reconstructed ancient wooden buildings were to be given active roles in interpretation of historic settlement (Weise, 2018, p. 124) and the economic aspect makes interaction between preservation of authenticity and tourism needs being more beneficial for the former one (Figure 10.14 – 10.19).

Figure 10.14. "A reinvention of tradition" People's nostalgia for their heritage and strengthening of national and ethnic identity; Tianjin`s Ancient Culture Street (Guwenhua Jie), a pedestrian pathway complex dotted with temple gates and kiosks on the west bank of the Hai River, Tianjin (photo T. Tomaszek, 2019).

Figure 10.15. The Old Town of Lijiang, Yunnan Province. It is one of the most famous clusters of traditional architecture of Naxi people, an ethnic group inhabited this spot from the Yuan Dynasty (1271–1368). After it was inscribed on the UNESCO World Heritage List in 1997, many changes have occurred on this historic site, for example the growing commercialization and outflow of original residents to outside the city center resulting the lost character of "living heritage" as well as changing its architectural character by implementing the reconstructions of "the old tradition" (photo T. Tomaszek, 2019).

Figure 10.16. The Old Town of Lijiang, Yunnan Province. After it was inscribed on the UNESCO World Heritage List in 1997, it lost its architectural charm as well as "living heritage" character (photo T. Tomaszek, 2019).

Figure 10.17. The Old Town of Lijiang, Yunnan Province. After it was inscribed on the UNESCO World Heritage List in 1997, it lost its architectural charm as well as "living heritage" character (photo T. Tomaszek, 2019).

Figure 10.18. The Old Town of Lijiang, Yunnan Province. After it was inscribed on the UNESCO World Heritage List in 1997, it lost its architectural charm as well as "living heritage" character (photo T. Tomaszek, 2019).

Figure 10.19. The Old Town of Lijiang, Yunnan Province. After it was inscribed on the UNESCO World Heritage List in 1997, it lost its architectural charm as well as "living heritage" character (photo T. Tomaszek, 2019).

More recently, questions have arisen on the authenticity of various ongoing trends, especially in respect to changes to the built heritage within the buffer zones (Figure 10.20 – 10.23). It becomes common to reconstruct only the fronts of the buildings, the rest keeping in the contemporary design. This has brought a new trend that could be called "facadism", whereby a lot of buildings based on reconstruction using reinforced concrete with imitated traditional facades are subsidized (Weise, 2018, p. 128).

The situation with protection of heritage in China changed drastically in the beginning of 21st century. Due to massive economic development and thus the need for a new approach to urban renewal and renovation. Local authorities officially started programs intended to maintain typical historic vernacular architecture and neighborhoods of individual regions. It was mostly meant for the purpose of tourism oriented utilization, and in consequence, in the case of many historic villages, the local stakeholders have even handed over the management of the whole historic site to tourism companies. The actions, which were focused on the manner of use of the protected buildings and their setting as a stage for performance (even if it was meant as historic performance) has brought serious changes to the original life of the local people in historic places and serious conflicts have arisen between traditional lifestyle and the introduced a narrative of an intended historical nature. Therefore, the cultural value of historic sites contained in the traditional crafts and way of life have rapidly begun to erode by modern tourism (Zhou, 2018, p. 98).

Figure 10.20. Imitation of a traditional type of wooden building in the contemporary materials/ design as the issue in maintaining authenticity of wooden architecture tradition; Da Luoshui Village on the Lugu lakeshore, Yunnan Province. This village is home to the Mosuo ethnic minority group famous for its matriarchal society and their wooden (log) architecture (photo T. Tomaszek, 2019).

The highly problematic case remains of the officially conducted preservation efforts by conservation professionals. Of the substantial amount of sites which present living heritage qualities, historical wooden structures are treated as being of far greater value than the rituals and religious rites for which these structures are erected. Though in reality, by nature

Figure 10.21. Imitation of a traditional type of wooden building in the contemporary materials/ design as the issue in maintaining authenticity of wooden architecture tradition; Da Luoshui Village on the Lugu lakeshore, Yunnan Province. This village is home to the Mosuo ethnic minority group famous for its matriarchal society and their wooden (log) architecture (photo T. Tomaszek, 2019).

Figure 10.22. Imitation of a traditional type of wooden building in the contemporary materials/ design as the issue in maintaining authenticity of wooden architecture tradition; Da Luoshui Village on the Lugu lakeshore, Yunnan Province. This village is home to the Mosuo ethnic minority group famous for its matriarchal society and their wooden (log) architecture (photo T. Tomaszek, 2019).

Figure 10.23. Imitation of a traditional type of wooden building in the contemporary materials/design as the issue in maintaining authenticity of wooden architecture tradition; Da Luoshui Village on the Lugu lakeshore, Yunnan Province. This village is home to the Mosuo ethnic minority group famous for its matriarchal society and their wooden (log) architecture (photo T. Tomaszek, 2019).

these structures are essentially the abode of religious rituals and need to be perceived in that way. The problem becomes even more dramatic, when no traditional rituals are permitted, and permission for any such use has to be authorized by the protected areas' management authority. For the local community, the act of protection, followed often by nomination to the list of protected monuments, results in a change of ownership of the sacred site from the community to the official government body. As a consequence, local community use of the site often is seriously restricted. While at the same time, the site becomes open to tourism and plays the role of a stage for an invented narration. For local communities, the concept of a "protected area" or "monument" has often resulted in fences, a denial of access, and desecration of sacred places through tourism (Nyathi; Ndiwini, 2005, p. 60-63).

In The China Principles ["Principles for the Conservation of Heritage Sites in China"], which is an important document containing the regulations for preservation of monuments, proposes a list of ten basic principles that should be followed for the interpretation of notion of authenticity in China. These principles refer to the original site, the original conservation state, and the original character of the setting (Zhou, 2018, p. 98). Particularly worthy of mention are the following postulates:

"2.2.1 Physical remains must be in their historic condition. This includes a site's condition as it was originally created, its condition after undergoing repeated adaptation throughout history, or its condition as a result of deterioration or damage over a long period.

2.2.2 Large complexes of buildings or historic precincts within villages and townships should retain their overall historic appearance. Modern additions, alterations, or loss should constitute only a small proportion of a site.

2.2.5 Recent imitations of historic landscapes that use an historical name or borrow the name of a heritage site are not to be considered heritage sites" (ICOMOS China, 2000, p. 15).

Later, in the revised version of the China Principles (ICOMOS China, 2015 (b)), an even a more nuanced set of guidelines regarding authenticity and the possible problems with the interpretation were included. Important statements include: "Authenticity resides in the original materials, workmanship and design of a site and its setting, as well as in its historical, cultural, and social characteristics and qualities. Respecting these aspects through conservation retains authenticity. The continuation of long-established cultural traditions associated with a particular site is also a means of retaining its authenticity" (ICOMOS China, 2015 (b), p. 67). [...] "The values of a heritage site should be presented and interpreted in an authentic, integrated, and accurate manner, and be based on comprehensive, in-depth research. Inaccurate interpretation of a site's history and culture must be avoided. Presentation and interpretation should meet the needs of various audiences and employ methods that are easily understandable" (ICOMOS China, 2015 (b), p. 100). [...] "Presentation should, as far as possible, provide a complete and accurate vehicle for the interpretation of the heritage site. Interpretation should be based on in-depth research without subjective conjecture. When there is not one definitive conclusion to research results, alternative explanations should be provided" (ICOMOS China, 2015 (b), p. 101).

Apart from above mentioned issues regarding interpretation, particularly when applied to the authenticity context, there is also a danger laying in the excessive respect to traditional crafts and techniques, prioritizing them over the value of the original fabric. It may often result in the replacement, partly or wholly of the original fabric with new, even without verifying whether the original parts have really reached a condition in which they failed to function or could continue. This is a particularly unsuitable approach towards timber monuments, as in which case this practice could only be an early and arbitrary termination of the authenticity of the material (Zhan, 2018, p. 91). It is important to underline, that the authenticity of material elements may be related directly to the respect of various traditions, beliefs and emotions and respect for legitimacy and necessity of social and realistic functions and spiritual needs (Zhan, 2018, p. 93), thus the value of the original material should in any case also be properly recognized. It could be said finally, the protection of the authenticity of a site means to protect its holistic truthfulness, including the historic, cultural as well as social information. At the same time, protection of a monument must combine conservation efforts with sustainable development. Which at all times must be connected to constant change as the nature of its reality.

Technical solutions in the conservation of wooden architecture in China and the question of authenticity

ABSTRACT: The degradation of wooden structures is difficult to prevent and has necessitated a cyclical process of maintenance and conservation. In this chapter the damage most common to historic wooden structures in China is described, as well as its causes. It is followed by a description of the various methods of reinforcement interventions for degraded wooden historic structures that are in use in China today. These reinforcement methods are in place due the specifics of Chinese traditional wooden structures coming from the usage of a construction based on a whole-beam frame. This part is supplemented with a discussion on authenticity of historical wooden structures in China in the view of the applied technical solutions, as well as the use of new engineered wood products for construction.

Buildings made of wood are much more perishable compared to buildings erected from permanent materials. They are easily destroyed by fire and the ravages of weather and time, and particularly sensitive to wars or natural cataclysms raging through historical times. As it is noted in Article 4 of *Qufu Declaration*, the main disadvantages of Chinese wood-framed buildings is that: "they are easily decayed, wormed and burned. Decay is the most important problem. The most perishable part of the timber frame is the wooden pillars buried in the walls and roofing wood beams, (...) The rotted root can easily cause the sinking of the column which will then cause the upper frame to deform, allowing the roof to leak. The result is that the leaking roof will increase the risk of damage to the timber frame" (Luo, et al., 2005).

Apart from typical problems like decay on the bottom of columns due to long-term exposure to humidity, the most common damage to historic wooden structures are the result of exertion of different forces. As proposed by Chen, and also Que, Li, Zhang, Yuan and Pan, they can be divided into three groups:

1) Component deformation under long lasting compression – for example, column yielding such as splitting of column tips, splitting along the grain on the body of columns, bending and splitting of girders and square beams between column and Dou-gong brackets as well as breaking and splitting of subcomponents of Dou-gong brackets.
2) Components deformation under tension – for example, square beams through the columns on the top or the bottom usually get adrift or break off at the mortise and tenon joints.
3) Components deformation under shear strength – for example, the mortise and tenon joints tend to detach under long-term shearing action (Chen, 2003); (Ze-li, et al., 2017, p. 215).

Degradation of wooden structures is difficult to prevent and has necessitated a cyclical process of maintenance and conservation. A well-known and often practiced method was, for example, the replacement of degraded elements or the reconstruction of damaged elements. This process itself, resulted in a set of customs and technical solutions that can be defined as the natural cycle of a wooden building. Thus, the use of these solutions, often sanctioned by millennia, fits in the horizon of the authenticity of the cultural tradition of wooden architecture in China.

The situation changed significantly at the end of the 20th century, when China opened up to patterns of different conservation methodologies permeating from Western culture and began to use new technologies to preserve the heritage of wooden architecture (Jin, 2019, p. 83). The solutions used, and the materials applied, put into question their correctness in terms of both the continuation of tradition and the problem of violating the authenticity of the structures in question.

In today's China, there are various methods of reinforcement interventions in usage for degraded wooden historic structures, wherein all are due the specifics of traditional Chinese wooden structures coming from usage of construction based on the whole-beam frame (excluding log structures). As proposed by Que, Li, Zhang, Yuan and Pan, these reinforcement interventions have different stages: (1) major repair of the structure, (2) restoration with external support, and (3) overall reinforcement (Ze-li, et al., 2017, p. 215). The first stage includes disassembling the wooden frame partially or completely, then repairing or replacing the damaged components and reassembling it while also reinforcing the structure; the second stage is adding a form of external support and restoring the tilted, twisted or detached components while reinforcing the structure without disassembling the frame; and the last is the direct reinforcement of the whole structure with minimum structural deformation.

Accordingly, it is possible to define the three most common types of reinforcement methods used in China today (Ze-li, et al., 2017, p. 215-216). They are as follows:

1) Reinforcement by complete or partial exchange of damaged components. The minimal intervention is the execution of patching, wherein the patch can be made when the splits or degradation of components like beams and columns are quite slight (Ze-li, et al., 2017, p. 215). It is common to use wooden powder and waterproof adhesives to achieve results. When the rotten part of a component such as a column takes up more than a quarter of the height, it is recommended to execute a partial exchange in that section. The rotten section of the column is then replaced with new materials, a matching wood with similar characteristics. The area where the exchanged part has been introduced is often also reinforced with a semi-tenon and an iron hoop. When the damaged depth of a component, for example a beam, takes up more than a third of its height and all of its sides, it is recommended that after exchanging the degraded part of the component, to use a clamp as an additional reinforcement method around the newly introduced part. The most intrusive and advance intervention is executed when the damaged depth of the component takes up more than three-fifths. In this case, the most appropriate solution is a complete exchange of the element.
2) Mechanical reinforcement. Most commonly used is an additional support using iron elements. And so, for example, to reinforce beams and columns or to connect joints between beams and columns, flat iron bars are often used. This is because the introduced flat iron bars can improve the mechanical properties of components, especially by bearing part of any forces (like tensile, compressing, bending or shearing). Tensile iron bars can be also introduced as load-bearing components. They are used in cases of more advanced deformation of elements, for example when deflection of beams and square pillars transcends the normal limitation or the load-bearing capacity is insufficient, They are also used if any splits threatening stabilization are found.
3) Chemical reinforcement. Since the 1970's, it was very popular to use an unsaturated polyester resin filler in the restoration of historic buildings in China (Ze-li, et al., 2017, p. 216). Different chemicals were (and still are) used, to introduce components by the filling, soaking, patching or painting. It is done in an attempt to improve the strength of the damaged wood and enhance the stability and anti-fungi capacity. Chemical reinforcement was used commonly in many important wooden structures in China. It can be found, for example, in the main palace of Nanchan Temple in Wutai, Shanxi province, coming from 782 AD. During its restoration in 1974, the splits of two main beams were filled with

epoxy resin. The spots then were strengthened additionally by the fixing iron hoops on them. An other case worth mentioning, is the main palace of Baoguo Temple in Ningbo, Zhejiang province, dated 1013 AD. During the restoration works in 1975 (which was done without disassembling the wooden frame), the columns destroyed by termites, were filled with chemicals and wrapped in fibre-reinforced plastic (FRP). As reported by Chun, Yu and Pan, there were made some calculation of costs, and according to them, despite the high expenses of the chemical used, about 30% of the whole budget was saved as disassembling of the whole frame was avoided (Chun, et al., 2013, p. 48).

The three reinforcement methods described above, show the process of changes in use of conservation solutions, when traditional methods began to be abandoned and the use of modern materials began. The introduction of reinforcing metal elements, although considered for many reasons as inappropriate (for example, exposure to moisture condensation at the contact point of metal with wood), is probably a less drastic application than the implementation of plastic fabric in the structure of wood. Apart from the aspect of the loss of authenticity of the material, the issue of cultural and religious traditions of China, where cyclicality and repetition included in wooden structure were their core, remains an important problem.

Apart from the new strengthening methods in modern China, new engineering wood materials for construction of structures with traditional characteristics as well as reconstruction of historical buildings are also commonly used. Among them particularly popular is glued-laminated timber (glulam). This fabric is considered, from the point of modern industry and the building market, to show all kinds of advantages, such as natural wood texture, high quality of anti-corrosion, high usage and stable physical properties. Additionally, as underlined by some authors, it is also characterized by great plasticity and special expressive ability that can rival steel structure (Ze-li, et al., 2017, p. 209). As an example of using glulam for "precise" reconstruction of historical buildings could be mentioned the case of Xiangji Temple in Hangzhou, Zhejiang province, in south-eastern China. This historically famous temple was first built in 1016 AD, the original building was destroyed by fire. It was then rebuilt in 2010 AD. For the reconstruction of the architectural complex, different methods and materials were used. And so the monastery, the guest house and the dorm rooms were erected using wood (there were built as the log structures), while the major structures of the temple's bell tower and drum tower, as well for the so called Kinnara hall were reconstructed using mainly steel. The rest of the structures, including the main hall, were at same time reconstructed using glulam. It brought possibility of some structural modification to the traditional design of the temple, as using glulam combined with a traditional roof allowed to create a column-free room inside of the hall (Ren; Liu, 2012, p. 86-88). The decision of such a reconstruction remains a matter of serious debate, even some of authors believe that "with the traditional multi-overhanging eaves, the main hall demonstrates a splendid momentum as well as openness and brightness" (Ze-li, et al., 2017, p. 209).

Less arguable seems to be the use of glulam for construction of buildings imitating the traditional structures, but at the same time representing its own era. Let's say that the building only refers to the historical structure (historic construction solutions or historic shape), but has a modern design and aesthetic solutions. This is for example a case of the White Lotus Lore Temple in Shanghai, where the main structure of Buddha hall was made of glulam and structured with a grid structure to make a column-free indoor space. With its large overhanging eaves and mild-slope roof, the hall was totally built according to the proportion of buildings in the Tang Dynasty. In this case, however, we are not dealing with heritage, but what is achieved, is just a modern building modelled on traditional solutions.

Other examples of using glulam for architectural structures following traditional designs (not being the "precise" reconstruction) and traditional construction methods are the Yu Xi Temple Tower in Chao Mountain in Hangzhou, Zhejiang province, in south-eastern

China and the Laojun tower in Qingcheng Mountain, Sichuan province, in south-western China. The first was built entirely of glulam as a pavilion-style tower on an octagonal layout with a pillar in the centre. The second one, after reconstruction being a type of imitation of a lost object, has the first and second floors made of concrete and glulam and from the third through ninth floors is made of glulam mixed with Douglas fir (Ze-li, et al., 2017, p. 210).

It is worth mentioning, that due to the specifics of Chinese culture even the construction of new palaces and temples in the country involve some kind of repetition of architectural patterns of pre-existing historic buildings or a direct reference to classic elements. However, these days a traditional solution of wooden structures are more frequently combined with modern construction. So, for example, this situation is the reason wooden structures are being combined with glass, concrete and steel. What is more, classical mortise and tenon joints are combined with metal joints and adhesives. This brings more flexibility in wooden structure design, but also raises the problem of losing the identity of Chinese traditional building and its formal as well as material authenticity.

However, as pointed out by some, contemporary modifications to traditional wooden building solutions come with many practical advantages, as they allow for example to extend the building scale. This is in the case of the above mentioned White Lotus Lore Temple, in which a wooden structure adopted a steel system such as the grid structure in and also Kai Yuan Temple, where the truss system was introduced. Some people postulate, that because this structural innovation and optimization, as well as advancement of construction techniques, contemporary wooden buildings, let's call them "contemporary traditional wooden structures", will become more perfect in structural logic, creativity and details. But can it hold the meaning of being traditional, when changing its historical, spiritual, social, esthetical and so on authenticity at almost all levels? Might therefore authenticity stand here to maintain specificity of cultural design and pattern with all its dynamics in the progress of historical time?

Final remarks

ABSTRACT: The term "authenticity" has become widely discussed and thus essential for contemporary theory as well the practice of architectural heritage conservation. After it was introduced to a Chinese context, "authenticity" has brought particular uncertainties and tensions into conservation discourse. It was mostly due to the interpretation of its notion followed an "international" understanding. At the same time, the particular approach towards describing authenticity in a Chinese context, as well as attempts of its applicability in the conservation practice in China, further resulted in bringing many doubts regarding the way how the Chinese government implements heritage conservation under its own interpretation of Chinese, as well international heritage guidelines. During the following decades authenticity was continuously interpreted, implemented, and adjusted to a Chinese context on the basis of its links with Chinese culture, history, and institutions. This chapter summarizes discussion on authenticity and conservation of wooden architecture in China, at the same time giving some overview of the spiritual and cultural values that are challenging the further discussion on this subject.

After being introduced in the 1980s, the term authenticity became widely discussed and thus essential for contemporary theory as well the practice of architectural heritage conservation. Shortly after being borrowed from international legislation and introduced to Chinese context, the term and its interpretation following "international" understanding of the notion, brought particular uncertainties and tensions into conservation discourse in China (Zhu, 2017, p. 189). At the same time, the particular approach towards authenticity in a Chinese context, as well as attempts of its applicability in the conservation practice in China, further resulted in bringing many doubts regarding the way how the Chinese government implements heritage conservation under its own interpretation of Chinese, as well international heritage guidelines.

There were several attempts to translate the meaning of authenticity to Chinese context. Yet, at the beginning, the translation prevailed as *yuan zhenxing* (原真性) which means both originality and truthfulness. Thus, it brought the idea of originality being applied to restoration projects in Chinese context. All the attempts to define authenticity were then furthered in issued by China ICOMOS (2000) the *Principles for the Conservation of Heritage Sites in China* (Zhou, 2018, p. 97). After the year 2000, the Chinese conservation practice began to shift its focus from mostly before relic-based to now more heritage-based (treating heritage as a whole system, combining tangible as well intangible aspects). It was already marked in the early version of China Principles that "The aim of conservation is to preserve the authenticity of all the elements of the entire heritage site and to retain for the future its historic information and all its values". This notion of authenticity was still reflected in the Qufu Declaration from 2005, where attention was placed on original style, original material, original structure and original craft, also possessing the scientific, artistic as well as historic value.

Following Liang Sicheng's ideas, and his attempts to define the original state or original form, in the beginning of 20th century it was (and still in many cases it is) interpreted as the state when a monument was first completed or in its prime. As a result, it comes to the practice of the so-called authoritative restoration observable in China up to the present, which promoted the elimination of other historical information that is considered non-original (Zhan, 2018, p. 91). Thus, it can be said that the original style is considered when the building is cleared of from historic layers. Additionally, the renovation style of the Soviet Union strongly influenced Chinese conservation practice, in many cases all attempts were directed towards pursuing the original state or the most glorious image of an historic building (Zhou, 2018, p. 97).

The introduction of the open-door policy in China, has resulted, amongst others, in wide ranging involvement with the international heritage conservation movements. Thus, also to the adoption of concepts emerging from the international arena. Therefore, it has been slowly but surely built into the Chinese context step by step and the significance of "the old" and the "value of history" has been recognized. Having been influenced by years of a Maoist perspective on cultural heritage, it was required to rethink and redefine this borrowed idea of "the old". However, the omnipresent Liang postulate of "restoring the old as it was", took the key position in understanding heritage "authenticity", particularly in the case of historic architecture preservation (Zhu, 2017, p. 193). At the same time, the meaning of "the old" has undergone a significant modification, and from understanding it as term implying something backward and hidebound, changed to an appreciation of its own tradition, culture, heritage and national identity. However, this understanding of "the old" does not always comply with the local practices in present day China, which practices show quite a different notion of this term. All over the country, we can come across some creations of "the old", manifested by "antique" products, particularly executed as "archaized buildings", which are imitate historic buildings. It comes as a direct response to the Chinese people's quest for authenticity and the visitors' imaginativeness that authorize this so common practice of "restoring the old" in this case represented by recreation. It can be almost indisputably be said, that the need for this practice arises from the people's nostalgia for their heritage, which appeared in the course of rapid modernization and industrialization which has been happening in China over the past few decades.

While from the point of western conservation theory, reconstructions maybe treated as serving two important functions: experimental research and interpretation (ICOMOS Charter for the Protection and Management of the Archaeological Heritage, (ICOMOS, 1990, Article 7), in Chinese context shows the nature of time and spiritual value of continuity. That is why it is so difficult to shift to Chinese tradition. Suggestions of western theories which basically question of historic time and authenticity in relation to the original object. All of it brings us to the edge of importance of the intangible part of heritage and it makes us aware of the difficulties in transferring the Western understanding of tradition, religiosity and thus the protection of cultural heritage onto the Chinese space of mentality and spirituality.

It can be said without any doubt, that wooden architecture symbolizes the spiritual and material well-being of Chinese civilization. Among wooden structures, the most significant to stand out are the wooden temples, being a direct expression of the most important intangible assets. The protection of this heritage therefore, is vital for continuity of the culture and spiritual tradition. It needs to be remembered, that the majority of wooden temples in China are living religious heritage, thus attempts to raise certain issues with regard to the application of the concept of authenticity must be understood in the context of living religious heritage. Religious structures are specifically living objects of architecture, particularly wooden Baddish architecture in China, in case of which, circularity and thus continuity it is the backbone of the spiritual meanings contained in them. "The living monuments require the forbearance of both devotee and specialist. Either extreme can produce distortion and eccentricity. On the one hand, the restoration or rebuilding concept of pilgrims and peasants can move monuments out of their own balance and replace such sites of hallowed dignity with monstrosities of a previous generation. On the other hand, the pure policy maker may be unmindful of the necessary balance and harmony that should exist between science and religion, and thereby make living and religious monuments no more than specimens of scientific interest" (Silva, 1983, p. 43). The concept of 'Living Heritage' is naturally connected to 'communities' and the 'continuity' of traditions and practices. Thus, it is 'continuity' that should be considered as a key feature that characterizes living heritage, whereas 'use for which it was originally intended' is the essential factor of defining 'continuity'. It should be noted, however, that this 'use' required for 'continuity' should not be confused with the notion of 'use' as some form of functioning or use that heritage places have for the present society. The 'use' referred to here is an idea that determines the authentic purpose of the building, and thus can be treated as determining the scope of authenticity when evaluating a given object (Wijesuriya, 2018, p. 47).

Chinese wooden historic architecture in its physical form represents tangible heritage, but its intangible value represented in materialization of idea goes far beyond the spectrum of authenticity residing in fabric. In Chinese context, the preservation of the material remains of a wooden historic building is not necessarily paramount, and there is a centuries old tradition of continuous renewal and rebuilding, where the process of reforming itself embodies essential spiritual significance. As a consequence, an historic building in which some older elements are obscured by more recent treatments is still understood as authentic, as its materialization represents a revitalizing spiritual process with the continuing embodiment of enduring beliefs and practices valued by the community to which it belongs. This approach as a conservation practice, has ramifications for an alternate cosmological view and different concept of time. It comes from regenerative view of the relationships between the past and the present, which is circular and continuous (Wijesuriya, 2008, p. 74).

This particular concept of time as circularity endowed in the nature of reality is observed in Buddhist tradition, one of the basis forming the foundations of Chinese culture. It needs to be remembered, that this circularity is permanently inscribed in activities of Buddhist societies, especially religious activities. Therefore, the renewal process or the continuity of temples is undertaken for various reasons such as earning merit (Wijesuriya, 2005, p. 36). As a consequence, authenticity must be looked at in relation to the renewal, the use of traditions including craftsmanship, techniques and the materials. It should be particularly be connected to the traditional yet changing knowledge systems that sustained these places for centuries. It can be said therefore, that the Buddhist wooden built heritage is a combination of both the tangible and intangible aspects – interrelated and interdependent. The traditions of the societies keeping continuity alive has led to the survival of the monuments and in other ways wooden shrines have caused a continuation of spiritual patterns and meaning of beliefs. In this light, it needs to be also viewed, the conservation practice of temples that was known in China for millennia. It seems to be an essential aspect for understanding the continuity of the temple structures actualizing the living nature of spiritual message(Kalupahana, 1975).

As it was discussed before, in contemporary theory and practice of preservation of wooden heritage the issues do not only concern the presence, or the absence, of efforts to conserve, but also how to conserve with the primacy of authenticity (Zerrudo, 2018, p. 106). It is due to the search for adequate maintaining of the value and significance, which are primarily based on authenticity. Therefore, authenticity of wooden buildings in Chinese tradition need to be treated as residing in and beyond materiality, as it is holistically enriched by traditional knowledge systems and rituals (Fadriquela, 2013). Consequently, the notions of authenticity need to be seen as a range from the most tangible manifestation to the most intangible expression included in physical form or eternality of place (Zerrudo, 2018, p. 111). It shows somehow a holistic combination of tangible and intangible attributes to authenticity, lightening the obsession to be, or as close as possible to, authentic. Authenticity thus, can be understood as nature of things, which is nothing other than the state of truthfulness.

Sacred wooden architecture represents, in particular, the inseparable link between the spiritual and the material form. It is a particularly true statement regarding China, where wooden temples express complicated patterns of spiritual and religious complexes, embodied in the broader spectrum of cultural forms and processes that show the dynamic interrelationship between tangible and intangible. Spirit or tradition cannot thus be separated from form, design, layout, material and craftsmanship as it underpins them – it is from the spiritual belief that the temple or sacred object emerges. Without the spiritual value, the material value may have limited meaning (Sharma, 2018, p. 178).

At the same time it must be remembered, that traditionally a wooden building in China was constructed in accordance with ancient recommendations, thus it followed "an authentic" process of erecting it, including particular ceremonial processes and formulated practices. And so, an arrangement of buildings, their style, size and layout, was heavily influenced by Buddhism spirituality and Confucian as well as Taoism philosophy. The same was in the case of repairing the wooden building, in which ancient recommendation was to be followed in order to consider the repair to be accepted as properly undertaken (Feng, 2012, p. 76).

It is very likely that the tradition of conservation of wooden temples in China, in practice forming a specific mix of preservation, reconstruction and restoration, began alongside the construction of the earliest religious images and architectural structures that were meant to house them. With passing of time, these buildings were developed and changed multiple times, which has allowed them to continue over centuries until this day. The aim of the understood conservation appears to have two causes, firstly, assuring continuity of spiritual and religious activity and secondly, to guarantee the longevity of the artifacts or edifices used in this religious activity (Tiwari, 2009, p. 5). Therefore conservation, being the cycles of construction, destruction, renovation and/or reconstruction led to overlapping various layers of history, meanings and materials, forming what we call today, wooden heritage of sacred sites in China.

After the concept of authenticity was adopted by the Chinese government, it was continuously interpreted, implemented, and adjusted to Chinese context on the basis of its links with Chinese culture, history, and institutions. The necessity to address the complexity of the concept, was the reason for the creation of a set of "official conservation guidelines". Among them, two documents stand out as the most important, which are the Law of the People's Republic of China on the Protection of Cultural Relics (The Law of the People's Republic..., 2007) and the Principles for the Conservation of Heritage Sites in China. Somehow in the response to an ongoing discourse on methodology of heritage conservation in China. *The Declaration of Qufu* —a *"Consensus on the China-specific Conservation Theory and Practices of Historic Buildings"* was adopted in 2005. This particular document was developed with the intention to serve as a specific principle for architectural heritage conservation particularly in Chinese cultural context. As in the art of preserving ancient buildings, China's ancient craftsmen have accumulated a wealth of practical knowledge, this document was therefore based on the centuries-old experience of traditional Chinese heritage protection practices (Jin, 2019, p. 91).

These new "international" principles, when taken from the perspective of a "Chinese framework," are in reality some kind of concession to both Liang's assertion of the principle of "restoring the old as it was" and postulates raised by the Venice Charter. This particular issue, is especially applicable to the context of today's China. Where the rapid development has brought noticeable changes in social life, the removal of original communities from built heritage. The process of "muzeumification" which is widely noticeable around the country, where many wooden historic structures receive the status of "monument taken out of history". The problematic assumption of authenticity as maintenance of fabric and intangible values, is lacking therefore of an important dimension of this value – the living community, necessary for keeping the site in constant change resulting directly from the process of continuity.

The natural process of continuity and repetition of patterns, broken during the Cultural Revolution, manifests itself in today's China with an increased need to replicate historical buildings. At the same time, until recently, disregarded the architectural wooden heritage of ethnic minorities which is of particular interest now. The case of which, the process of cultural continuity has often, so far, been undisturbed. This recognition of the importance of its own heritage, could be seen an attempt for China to rebuild the broken link of continuity. This process also brings a direct response to the attempt of defining the semantic scope of the notion of authenticity, a foreign term, and a little automatically transferred from international conservation legislation. As a value in itself, authenticity relying on continuity may shift the meaning of preserving China's wooden architectural heritage from a preservation perspective focused on fabric to a value-focused approach. However, this type of static model cannot fully describe, not only the tradition of continuity of the architectural form in China, but also the concept of living heritage as a continuity of the original purpose of the building.

It needs to be stressed finally, that the traditional Chinese approach towards wooden architecture is diametrically opposed to western, particularly the European approach. While the first is more about the understanding of a culture that has been going through a continuous process of adaptation and change, the second, in contrary, rather than looking at changes in the structure over time is fixed upon very specific changes in specific time periods. And this fixed time is treated as a baseline for preservation in western tradition, but continuity naturally included in tangible as well as intangible aspects of cultural heritage does not necessarily provide justification for the fixed time in question (Weise, 2018, p. 125).

Mentioned above, the *Qufu Declaration* particularly prioritizes the methods of conservation of wooden structures that were used traditionally in China. Therefore, it allows us to increase our understanding of ancient construction methods, knowledge of which is essential for their proper preservation. The significance of ancient construction lies on not only the existence of a specific building in a certain historical period, but of the same importance, is the continuous renewal and circulation of specific architectural forms. Thus, the process of continuous reconstruction of China's ancient construction leads to natural metabolism and continuation, which is an essential method for the knowledge of wooden structures to be handed down to future generations (Jin, 2019, p. 96-99).

Figure 12.1. The guardian of the historic wooden building dusting the window sill; Forbidden City, Beijing (photo T. Tomaszek, 2018).

References

Agnew N., Demas M., Sullivan S., Altenburg K. 2004. *The Begetting of Charters: Genesis of the China Principles.* /in:/ Historic Environment 1/18 (September). pp.: 40–45.

Ahmed A. S. M., Huq N. E. 2018. *Authenticity: resynchronizing theory and practice – analysis of the decision-making process in the conservation arena of Bangladesh* /in:/ Wijesuriya G., Sweet J. (eds.). 2018. *Revisiting Authenticity in the Asian Context.* Conservation Forum Series. ICCROM, Rome. Vol. 2. pp.: 189–196.

Allen C. 1982. *A Mountain in Tibet: The Search for Mount Kailas and the Sources of the Great Rivers of Asia.* André Deutsch, London.

Ancient buildings in Wutai County-Nanchan Temple. Undated. Xinzhou City Cultural Relics Bureau. Xinzhou Net. (accessed: 29. 08. 2018), (in Chinese).

Andrews T., Buggey S. 2008. *Authenticity in Aboriginal Cultural Landscapes.* /in:/ Apt Bulletin: Journal of Preservation Technology. Vol.: 39. No. 2/3.

Anonymous. 1990. *Jixian History: Dule Temple.* Yuelu Publishing House, Beijing (in Chinese)

Anyon R. 1991. *Protecting the Past, Protecting the Present: Cultural resources and American Indians.* /in:/ Smith G.S., Ehrenhard J.E. (eds.). 1991. *Protecting the Past.* CRC Press, Boca Raton. pp.: 215–222.

Araoz G. 2011. *Preserving Heritage Places under a new Paradigm.* /in:/ Journal of Cultural Heritage and Sustainable Development. Vol.: 1(1). pp.: 55–60.

Araoz G. 2013. *Conservation Philosophy and its Development: Changing Understanding of Authenticity and Significance.* /in:/ Heritage and Society. Vol. 6. No 2. pp.: 144–154.

Assmann J., Czaplicka J. 1995. *Collective Memory and Cultural Identity.* /in:/ New German Critique, Vol. 65. pp.: 125–133.

Bramble C. 2003. *Architect's Guide to Feng Shui: Exploding the Myth.* Elsevier, Oxford.

Buswell E. R. 2004. *Encyclopedia of Buddhism: A-L.* Macmillan Library Reference.

Camaria P. 1996. *Kailash Manasarovara on the Rugged Road to Revelation.* Abhinav, New Delhi.

Chai J.Z. 1986. *Architectural form analysis of the East Palace of Foguang temple.* /in:/ Mt. Wutai Researches, Volume 1, pp.: 17–20 (in Chinese).

Chai Z. 1999. *Chai Zejun Gujianzhu Wenji.* Wenwu, Beijing (in Chinese).

Chang Q. 2009. *Debate of Authenticity in Historical Architectural Conservation.* /in:/ Time Architecture, Vol. 3. pp.: 118–121 (in Chinese).

Chapagain N. K., Tiwari S. R. 2018. *Challenges to the notion of authenticity in the context of temples in Kathmandu* Valley, *Nepal* /in:/ Wijesuriya G., Sweet J. (eds.). 2018. *Revisiting Authenticity in the Asian Context.* Conservation Forum Series. ICCROM, Rome. Vol. 2. pp.: 133–139.

Chen G.Y. 2003. *Study on material properties change of old wood of historic buildings and the effect on building deformation.* /in:/ Traditional Chinese Architecture Gardens, Volume 3. pp.: 49–52 (in Chinese).

Chen M. 1954. *Newly discovered ancient buildings in Shanxi Province in the past two years.* Reference materials for cultural relics. Cultural Relics Publishing House, Beijing. (in Chinese).

Chen M.D. 1966. *Yingxian Wooden Pagoda.* Cultural Relics Press, Beijing (in Chinese).

Chen Z.Y., Zhu E.C., Pan J.L. 2013. *Lateral structural performance of Yingxian Wooden Pagoda based on refined FE models.* /in:/ Journal of Building Structures, volume 34(9), pp.: 150–158 (in Chinese).

Chinese Academy of Architecture. 1982. Ancient Chinese architecture. China Building Industry Press. (accessed: 13. 10. 2019).

Chödje D. N. 2003. Paper presented at the Working Group Meeting on Restoration and Conservation of Monastic Heritage of the Himalayan Region, October 2003, New Delhi. UNESCO and NIRLAC in collaboration with the National Museum, New Delhi.

Chun Q., Yu M.Z., Pan J.W. 2013. *Research on damage characteristic and structural performance of the main hall of Baoguo Temple in Ningbo.* /in:/ Sciences of Conservation and Archaeology, volume 25(2). pp.: 45–51 (in Chinese).

Collection of Researches on the Protection of Shanxi Cultural Relics. 2011. Shanxi Ancient Building Conservation Research Institute. China Construction Industry Press. Beijing. (in Chinese).

Cooray N. 2012. *Current Issues of and Future Tasks for Conservation of Brick and Stone Stone Structures in Sri Lanka.* /in:/ Cultural Heritage Protection Cooperation Office, Asia-Pacific Cultural Centre for UNESCO, (ed.) Proceedings of the International Conference "Human Resource Development for Transmission of Traditional Skills – National Approaches and Their Application to Stone and Brick", 6–8 December 2011, Shanghai. Cultural Heritage Protection Office of Asia-Pacific Cultural Centre for UNESCO (ACCU). pp.: 109–135.

Cooray N. 2018. *Authenticity in the context of living religious heritage: the Sri Lankan experience* /in:/ Wijesuriya G., Sweet J. (eds.). 2018. *Revisiting Authenticity in the Asian Context.* Conservation Forum Series. ICCROM, Rome. Vol. 2. pp.: 153–162.

Cultural Heritage Policy Documents, Recommendations of the Madrid Conference (1904), Sixth International Congress of Architects. Undated. http://www.getty.edu/conservation/publications_resources/research_resources/charters/charter01.html. (accessed: 19. 11. 2019).

Dagens B. (trans.). 1993. *Mayamata: An Indian Treatise on Housing Architecture and Iconography.* Motilal Banarsidass Publishers, New Delhi.

Dagens B. (trans.). 1995. *Mayamata: An Indian Treatise on Housing Architecture and Iconography.* Sitaram Bhartia Institute of Science and Research, New Delhi.

De Vos A. 2018. *Authenticity in connection to traditional maintenance practices: sacred living heritage sites of Sri Lanka* /in:/ Wijesuriya G., Sweet J. (eds.). 2018. *Revisiting Authenticity in the Asian Context.* Conservation Forum Series. ICCROM, Rome. Vol. 2. pp.: 163–169.

Dellheim C. 1982. *The Face of the Past.* Cambridge University Press, Cambridge.

Department of Archaeology. 1904. *Epigraphia Zeylanica: Being Lithic And Other Inscriptions of Ceylon.* published for the Government of Ceylon by H. Frowde, London.

Department of Archaeology. 2007. *The Integrated Management Framework for the Kathmandu Valley World Heritage Site.* Department of Archaeology, Government of Nepal and UNESCO Kathmandu Office, Kathmandu. (in Nepali).

Dhumal R. 2018. *Authentic gold! Considerations on authenticity based on the gold plating at the World Heritage Site of Mahabodhi Temple at Bodh Gaya in India.* /in:/ Wijesuriya G., Sweet J. (eds.). 2018. *Revisiting Authenticity in the Asian Context.* Conservation Forum Series. ICCROM, Rome. Vol. 2. pp.: 181–188.

Dida G. 2003. *History of Ancient Chinese Architecture.* China Construction Industry Press, Beijing. pp.: 265–287. (in Chinese).

Dogra C. S. 2008. *The Broken Hemisphere.* /in:/ Outlook Magazine. no 3. November 2008. (also available at www.outlookindia.com//magazine/story/the-brokenhemisphere/238849).

Du L.M., Li H.W., Xue F., Qin D.Q. 2010. *The study of seismic behaviors of Yingxian Wooden Pagoda.* /in:/ China Civil Engineering Journal, volume 43, pp.: 363–370 (in Chinese).

Emerick K. 2003. *Use, value and significance in heritage management in Layton.* /in:/ Stone R., Thomas P.G. (eds.). 2003. *Destruction and conservation of cultural property.* London. pp.: 276–285.

Fadriquela C.S. 2013. *Kahoy: Wood in the Philippines (Wood and Its Uses from Pre-Hispanic to Spanish Colonial Philippines).* University of Santo Tomas Publishing House. Manila. Vol. 1.

Fairbank W. 1994. *Liang and Lin: Partners in Exploring China's Architectural Past.* University of Pennsylvania, Philadelphia.

Fan C.M., Wang L.A., Pan J.L. 2006. *Measures to control seasoning checks of wood used for the repair of Yingxian Wooden Pagoda, Shanxi Province.* /in:/ Journal of Beijing Forestry University, volume 2(1), pp.: 98–102 (in Chinese).

Feilden B.M. 1979. *An Introduction to Conservation of Cultural Property.* UNESCO, Paris.

Feilden B.M. 1982. *Conservation of Historic Buildings*. Butterworths, Oxford.
Feng J. 2012. *Chinese Architecture and Metaphor: Song Culture in the Yingzao Fashi Building Manual*. University of Hawai'i Press, Honolulu.
Foguang Temple 2005-2009 Progress Report. 2009. Global Heritage Fund, in partnership with Shanxi Bureau of Cultural Relics and the Shanxi Institute of Ancient Architectural Conservation. Shanxi Institute. China. http://ghn.globalheritagefund.com/uploads/documents/document_1448.pdf. (accessed: 14. 08. 2021)
Forsyth W.A. 1913. *The repair of ancient buildings*. /in:/ The Architectural Journal. 3rd Series, Vol. 21. pp.: 109–137.
Galla A. 1994. *Authenticity: Rethinking Heritage Diversity in a Pluralistic Framework*. /in:/ Larsen K.E. (ed.). Nara conference on authenticity in relation to the World Heritage Convention. Conference de Nara sur l'authenticite dans le cadre de la Convention du Patrimoine Mondial, 1–6 November 1994, Nara, Japan. UNESCO World Heritage Centre, Paris and Agency for Cultural Affairs of Japan, Tokyo. pp.: 315–322.
Gao J. 1984. *Aesthetics and Art: Traditional and Contemporary China in a Comparative Perspective*, Springer
Geiger W. 1912. *The Mahavamsa or the Great Chronicle of Ceylon*. Oxford University Press, London.
Gethin R. 1998. *The Foundations of Buddhism*. Oxford University Press, Oxford.
Gutschow N., Vasavada R. 2011. *Jain Architectural Heritage*. www.onlinefilm.org/en_EN/film/56456#!language/en_EN. (accessed: 20. 07. 2019).
Hao C.R. 2004. *Perspective of wood buildings in China view from the development of Chinese and western wood buildings* (thesis). Tsinghua University, Beijing (in Chinese).
Hou W.D. 2014. *Disputes and debates: conservation philosophy of the Yingxian Wooden Pagoda*. /in:/ World Architecture, volume 12; pp.: 38–41. DOI: 10.16414/j.wa.2014.12.005
Howard F. A. (et al.). 2006. *Chinese Sculpture*. Yale University Press, New Haven.
Huang H. 2012. *Background Analysis of the Birth of Legalization of Antiques Preservation (1911–1930)*. /in:/ Journal of the National History Academy, no. 2012. pp.: 22–26.
ICOMOS China. 2000. *Principles for the Conservation of Heritage Sites in China*. Burwood. http://australia.icomos.org/wp-content/uploads/Principles-for-the-Conservation-of-Heritage-Sites-in-China.pdf. (accessed: 5. 02. 2020), (in Chinese and English).
ICOMOS China. 2004. *Principles for the Conservation of Heritage Sites in China*. Issued by China ICOMOS, approved by the State Administration of Cultural Heritage, China ICOMOS, and The Getty Conservation Institute. 2004. Translated by Agnew N. and Demas M. First printed in 2002, second printing with revision 2004. Ventura Printing, Los Angeles. http://www.getty.edu/conservation/publications_resources/pdf_publications/heritage_sites_china.html, (accessed 30 September 2019).
ICOMOS China. 2015 (a). *Principles for the Conservation of Heritage Sites in China*. Wenwu Press, Beijing.
ICOMOS China. 2015 (b). *Principles for the Conservation of Heritage Sites in China* (revised edition). Paris. http://openarchive.icomos.org/1650/1/china_prin_heritage_sites_2015.pdf. (accessed: 05. 02. 2020), (in Chinese and English).
ICOMOS. 1965. *The Venice Charter* (International Charter for the Conservation and Restoration of Monuments and Sites), Approved by the Second International Congress of Architects and Technicians of Historic Monuments in Venice from 25 to 31 May 1964, adopted by ICOMOS in 1965. http://www.icomos.org/charters/venice_e.pdf. (accessed 15 January 2019).
ICOMOS. 1990. *Charter for the Protection and Management of the Archaeological Heritage*. Paris. http://www.icomos.org/images/DOCUMENTS/Charters/arch_e.pdf. (accessed: 04. 12. 2018).
ICOMOS. 1994. *Nara Document on Authenticity*. Nara, Japan. http://whc.unesco.org/document/116018. (accessed: 23. 09. 2020).
ICOMOS. 1999 (a). *Charter on the Built Vernacular Heritage*. Paris. http://www.icomos.org/images/DOCUMENTS/Charters/vernacular_e.pdf. (accessed: 19. 9. 2019).

ICOMOS. 1999 (b). *International Cultural Tourism Charter: Managing Tourism at Places of Heritage Significance*. Paris. http://www.icomos.org/charters/tourism_e.pdf (accessed: 07. 04. 2019).

ICOMOS. 1999 (c). *Principles for the Preservation of Historic Timber Structures*, Adopted by ICOMOS at the twelfth General Assembly in Mexico, ICOMOS October 1999; http://www.icomos.org/iiwc/charter-eng.htm. (accessed 20 March 2019).

ICOMOS. 2003 (a). *Principles for the Analysis, Conservation and Structural Restoration of Architectural Heritage*. https://www.icomos.org/charters/structures_e.pdf (accessed: 30. 07. 2020).

ICOMOS. 2003 (b). *Principles for the Preservation and Conservation/Restoration of Wall Paintings*. https://www.icomos.org/en/what-we-do/focus/179-articles-en-francais/ressources/charters-and-standards/166-icomosprinciples-for-the-preservation-and conservationrestoration-ofwall- paintings. (accessed: 13. 08. 2020).

ICOMOS. 2007. *Beijing Document on the Protection and Restoration of Built Heritage in East Asia*. http://orcp.hustoj.com/2016/06/04/beijingdocument-on-the-conservation-and-restoration-of-historic-buildingsin-east-asia2007. (accessed: 15. 06. 2020).

ICOMOS. 2017. *ICOMOS Ename Charter for the Interpretation of Cultural Heritage Sites*. Paris. http://www.enamecharter.org. (accessed: 24. 11. 2018).

Jianwei W., Tieying L., Shiwen L. 2002. *Measurement and analysis of dynamic characteristics of Guanyin Pavilion of Dule Temple*. /in:/ Journal of Taiyuan University of Technology. Vol. 7/ 2002. (in Chinese).

Jin X. 2019. *The Development of Protection Activities for Chinese Architectural Heritage and Its Unique Conservation Concepts*. /in:/ Wrocławskie Studia Erazmiańskie, Volume XIII – Orient far and near, Sadowski M. (ed.). Uniwersytet Wrocławski, Wrocław. pp.: 81–104.

Jokilehto J. 1986. *A History of Architectural Conservation. The Contribution of English, French, German and Italian Thought towards an International Approach to the Conservation of Cultural Property*. Institute of Advanced Architectural Studies, The University of York. http://www.iccrom.org/ifrcdn/pdf/ICCROM_05_HistoryofConservation00_en.pdf. (accessed: 14. 10. 2019).

Jokilehto J. 2004. *A History of Architectural Conservation*. Elsevier Butterworth-Heinemann, Oxford.

Jokilehto J. 2006. *Considerations on authenticity and integrity in world heritage context*. /in:/ City & Time. Vol. 2 (1), no. 1.

Jokilehto J., King J. 2001. *Authenticity and Conservation: Reflections on the Current State of Understanding*. /in:/ Saouma-Forero G. (ed.) *Authenticity and Integrity in an African Context*. UNESCO, Paris. pp.: 33–39.

Jugnu S. K. (ed. and trans.). 2013. *Great sage Atreya's Atreya-Tilakam: an Indian Treatise About the Principles of Image-Making*. Parimal Publications, Delhi.

Kalupahana D. 1975. *Causality: The Central Philosophy of Buddhism*. University of Hawaii Press, Honolulu.

Kawan S. S. 2015. *Heritage in all, heritage for all: Integrity overrides authenticity*. /in:/ Weise K. (ed.). 2015. *Revisiting Kathmandu Safeguarding Living Urban Heritage*. Proceedings of an international symposium Kathmandu Valley, 25–29 November 2013. Paris and Kathmandu, UNESCO and UNESCO Office in Kathmandu. pp.: 72–83.

Kim B. 1999. *A spirit discharged on this earth*. Isanggeonchuk, Seoul (in Korean).

Kim S. 2016. *Restoration Principles and Example of Maintaining Authenticity of Korean Architectural Wooden Heritage*. /in/ Cultural Heritage Protection Cooperation Office, Asia-Pacific Cultural Centre for UNESCO (ACCU). (eds.). Revisiting the Philosophy of Preserving Wooden Structures: Value of Wooden Structures in Asia and the Concept of Authenticity, 15–17 December 2015, Nara, Japan, pp.: 27–37. Cultural Heritage Protection Cooperation Office, Asia- Pacific Cultural Centre for UNESCO (ACCU), Nara.

Kim S. 2018. *Architectural heritage restoration and authenticity in Korea: a comparative analysis*. / in:/ Wijesuriya G., Sweet J. (eds.). 2018. *Revisiting Authenticity in the Asian Context*. Conservation Forum Series. ICCROM, Rome. Vol. 2. pp.: 65–75.

Knapp R. G. 2006. *Chinese Houses: The Architectural Heritage of a Nation*. Tuttle Publishing, Tokyo.

Kohrman M. 1998. *Technology and Gender: Fabrics of Power in Late Imperial China.* /in:/ American Anthropologist, no. 100. pp.: 236–236.

Kulatunga T.J. 1999. *Abhayagiri Vihara at Anuradhapura.* Ministry of Cultural and Religious Affairs, Central Cultural Fund, Colombo.

Lai G., Demas M., Agnew N. 2004. *Valuing the Past in China: The Seminal Influence of Liang Sicheng on Heritage Conservation.* /in:/ Orientations, Volume 35/2. pp.: 82–89.

Larsen K.E. (ed.). 1994. *Nara conference on authenticity in relation to the World Heritage Convention.* Proceedings, Nara, Japan, 1–6 November 1994. UNESCO World Heritage Centre, Agency for Cultural Affairs of Japan, ICCROM and ICOMOS. Trondheim. http://www.icomos.org/charters/nara-e.pdf. (accessed: 02. 05. 2020).

Larsen K.E., Marstein N. (eds.). 1994. ICOMOS International Wood Committee, 8th International Symposium. Kathmandu, Patan and Bhaktapur, Nepal 23 – 25 November 1992. ICOMOS with support from UNESCO and Directorate for Cultural Heritage – Norway. Tapir Forlag, Norway.

Law of the People's Republic of China on Protection of Cultural Relics. 2015. People's Republic of China. http://hk.lexiscn.com/law/law-english-1-2722291-T.html. (accessed: 15. 01. 2020), (online in Chinese and English).

Lee S. 2018. *Redefining authenticity in the Korean context: focusing on authenticity of form.* /in:/ Wijesuriya G., Sweet J. (eds.). 2018. *Revisiting Authenticity in the Asian Context.* Conservation Forum Series. ICCROM, Rome. Vol. 2. pp.: 55–64.

Li L., Tang L., Zhu H., Zhang H., Yang F., Wenmin Q. 2017. *Semantic 3D Modeling Based on CityGML for Ancient Chinese-Style Architectural Roofs of Digital Heritage.* /in:/ ISPRS International Journal of Geo-Information, no. 5.

Li T.Y., Qin H.M. 2005. *Structural analysis and repair of Yingxian Wooden Tower.* /in:/ Engineering Mechanics, Volume 22, pp.: 199–212 (in Chinese).

Liang S. 1932. *A Research on the Gate of Guanyin Pavilion of Dule Temple in Jixian County.* /in:/ Journal of China Architecture Society. Vol. 3. No. 2. (in Chinese).

Liang S. 1984 (a). *A Pictorial History of Chinese Architecture: A Study of the Development of its Structural System and the Evolution of its Types.* Fairbank W. (ed.). The MIT Press, Cambridge.

Liang S. 1984 (b). *Talking about the repair and restoration of cultural properties.* /in:/ China Architectural Press, Beijing. Volume IV of the Liang Sicheng Collection. pp.: 331–339 (in Chinese).

Liang S. 2001. *The Complete Albums of Liang Sicheng*, Vol. 9. China Architectural Industry Publisher, Beijing. pp.: 15–50 (in Chinese).

Lin N., Ran Z. 2002. *The Ancient Timber Members.* /in:/ Outlook Weekly, Volume 41. pp.: 52–53 (in Chinese).

Lin Z. 2008. *A History of the Society for Research in Chinese Architecture.* Baihua Literature and Art Publishing House, Tianjin

Linstrum D. 1991. *Principles of Conservation.* The York Gospel, University of York, York.

Little S., Eichman S. 2000. *Taoism and the Arts of China.* University of California Press.

Liu D.Z. 1984. *History of ancient Chinese architecture* (2nd ed.). China Architecture & Building Press, Beijing (in Chinese).

Liu X., Wang Z. Undated. *Nanchan Temple, the oldest wooden structure of the Tang Dynasty in China, is located in a corner of 'Treasure'.* Xinhua Net. (accessed: 27. 08. 2019), (in Chinese).

Liu Z. 2016. *The Case of Repatriating China's Cultural Objects*, Springer

Lowenthal D. 1996. *The Heritage Crusade and the Spoils of History.* Cambridge University Press, Cambridge.

Lowenthal D. 1999. *Authenticity: Rock of Faith or Quicksand Quagmire?.* /in:/ The Getty Conservation Institute Newsletter. Vol. 14, no. 3. pp.: 5–8.

Luo Z. 1999. *The Chinese Architectural Conservation and Restoration of Cultural Property.* /in:/ The World of Cultural Property 1/2. pp.: 19–30.

Luo Z., et al. 2005. *Qufu Declaration: Consensus on the theory and practice of architectural relics conservation with Chinese characteristics.* https://wenku.baidu.com/view/01611c43011ca300a6c3907b.html. (accessed: 13. 07. 2019).

Ma B.J. 2003. *The construction technology of China ancient wooden buildings* (2nd ed.). Science Press, Beijing (in Chinese).

Ma S. 1998. *Bogowie Chin*, Beijing Yanshan Publishing House, Beijing. pp.: 497–498. (in Chinese).

Majid Z. 2018. *An exploration into authenticity in archaeology for World Heritage nominations* /in:/ Wijesuriya G., Sweet J. (eds.). 2018. *Revisiting Authenticity in the Asian Context.* Conservation Forum Series. ICCROM, Rome. Vol. 2. pp.: 47–53.

Marshall J. 1990. *Conservation Manual. A Handbook for the Use of Archaeological Officers and Other Entrusted with the Care of Ancient Monuments.* Asian Educational Services, New Delhi. First published 1923.

Matunga H. 1994. *Wahi Tapu: Maori sacred sites.* /in:/ Carmichael D.L., Hubert J., Reeves B., et al. (eds). 1994. *Sacred sites, sacred places.* Routledge, London. pp.: 217–226.

Method for field test of natural durability of wood. 1993. National Technical Committee 41 on Timber of Standardization Administrator of China, GB/T 13942.2-92. China Standard Press, Beijing.

Millennium ancient temple. Dule Temple conducts fire rescue drills. Undated. Xinhuanet. (accessed: 14. 11. 2019), (in Chinese).

Mount Wutai (China), No 1279, UNESCO World Heritage Centre. Inscription form for UNESCO heritage list. https://whc.unesco.org/document/152245. (accessed: 17. 07. 2020).

Nara Document on Authenticity, Approved: Nara Conference on Authenticity in Relation to the World Heritage Convention, Nara, Japan, 1–6 November 1994. Proceedings. 1995. Larsen K. E. (ed.), XXI–XXV, Tapir Publishers, Trondheim.

Nara+20: On heritage practices, cultural values and the concept of authenticity. 2014. Agency for Cultural Affairs for Japan. Tokyo. www.japan-icomos.org/pdf/nara20_final_eng.pdf. (accessed: 10. 04. 2020).

Ndoro W. 2018. *Drawing parallels: authenticity in the African context – until lions learn to write, hunters will tell their history for them* /in:/ Wijesuriya G., Sweet J. (eds.). 2018. *Revisiting Authenticity in the Asian Context.* Conservation Forum Series. ICCROM, Rome. Vol. 2. pp.: 37–46.

Noga C.-K. 2008. *Religious Tourism and Pilgrimage Management: An International Perspective.* / in:/ The Professional Geographer. Vol. 60. No. 4. pp.: 589–591.

Nyathi P., Ndiwini B. 2005. *A Living Religious Shrine under Siege: The Njelelle Shrine/King Mzilikazi's Grave and Conflicting Demands on the Matopo Hills area of Zimbabwe.* /in:/ Stovel H., Stanley-Price N., Killick R. (eds.). 2005. *Conservation of Living Religious Heritage*, ICCROM Conservation Studies 3. ICCROM. Rome. pp.: 58–68.

ORCP (Open Repository on Cultural Property). 2017. *Law of the People's Republic of China on Protection of Cultural Relics* (2015 Amendment). Hangzhou. http://orcp.hustoj.com/2017/06/16/law-of-cultural-relics-2015-amendment. (accessed: 12. 10. 2019).

Pan G.X. (ed.). 2015. *A history of Chinese architecture* (7th ed.). Architecture &Building Press, Beijing (in Chinese).

Paranavitana S. 1945. *Administrative Report of the Department of Archaeology 1940–45.* Government Publication Bureau, Colombo.

Paranavitana S. 1947. *Protection of Ancient Monuments other than Those on Crown Land. A Pamphlet.* Government Publication Bureau, Colombo.

Peou H. 2018. *Safeguarding authenticity in the context of sustainable development at Angkor World Heritage Site: the role of the site's traditional hydraulic system* /in:/ Wijesuriya G., Sweet J. (eds.). 2018. *Revisiting Authenticity in the Asian Context.* Conservation Forum Series. ICCROM, Rome. Vol. 2. pp.: 197–206.

Perera H.R. 2016. *Buddhism in Sri Lanka: A Short History.* Buddhist Publication Society, Colombo.

Philippot P. 1976. *Historic Preservation: Philosophy, Criteria, Guidelines.* /in:/ Timmons S. (ed.). *Preservation and Conservation; Principles and Practices.* Proceedings of the North American International Regional Conference, 10–16 September 1972, Williamsburg, Virginia and Philadelphia, Pennsylvania. National Trust for Historic Preservation in the United States, Washington.

Philippot P. 1996. *Restoration from the Perspective of Humanities in Historical and Philosophical Issues.* /in:/ Stanley-Price N., Tally K. Jr., Vaccaro A.M. (eds.). 1996. *Historical and Philosophical Issues in the Conservation of Cultural Heritage.* Readings in Conservation Series, The Getty Conservation Institute, Los Angeles. pp.: 216–229.

Poshyanandana V. 2008. *Interpretation with Authenticity.* /in:/ Proceedings of International Symposium 2007 on Interpretation: From Monument to Living Heritage. Amarin Printing Group, Bangkok. pp.: 264–295.

Poshyanandana V. 2018. *The concept of authenticity in heritage conservation in Thailand: a case study of anastylosis.* /in:/ Wijesuriya G., Sweet J. (eds.). 2018. *Revisiting Authenticity in the Asian Context.* Conservation Forum Series. ICCROM, Rome. Vol. 2. pp.: 77–88.

Pregadio F. 2008. *The Encyclopedia of Taoism*, Psychology Press. Vol. 1.

Prince H. 1981. *Revival, restoration, preservation: changing views about antique landscape features.* /in:/ Lowenthal D., Binney M. (eds.). *Our Past Before Us: Why do we save it.* The Blackwell Press, London. pp.: 33–50.

Qin X. (ed.). 2004. *Dudong Wutaishan.* Shanxi People's Press, Taiyuan. (in Chinese).

Qufu Declaration. 2005. Consensus on the China-specific conservation theory and practices of historic buildings, adopted at the conclusion of a national conference on the Conservation theory and practice of historic buildings in China. Qufu.

Ratnyake P. B. 2018. *Authenticity in the Sri Lankan context: traditional maintenance systems, modern management systems, and present challenges* /in:/ Wijesuriya G., Sweet J. (eds.). 2018. *Revisiting Authenticity in the Asian Context.* Conservation Forum Series. ICCROM, Rome. Vol. 2. pp.: 141–152.

Recommendations of the Madrid Conference (1904), Sixth International Congress of Architects. 1909. Royal Institute of British Architects. The Architectural Journal, Being the Journal of the Royal Institute of British Architects (RIBA), Vol.: 11(3). pp.: 343–346. http://www.getty.edu/conservation/publications_resources/research_resources/charters/charter01.html. (accessed: 12. 04. 2020).

Ren L.Z., Liu J. 2012. *The application of glued laminated timber in the design of contemporary Buddhist architecture: rehabilitation of Hangzhou Xiangji Temple*, New Architecture. pp.: 86–89, DOI: 1000-3959 (2012) 05-0086-04.

Robson J. 2009. *Power of Place: The Religious Landscape of the Southern Sacred Peak (Nanyue) in Medieval China.* Harvard University Asia Center, Cambridge, Massachusetts.

Sada S. 1932. *Dura Temple in Jixun—the oldest existing wooden building and the largest statue in China.* /in:/ Study of Fine Arts, August 1932. (in Chinese).

Schaeffer K.R. 2011. *New Scholarship in Tibet, 1650–1700.* /in:/ Pollock S. (ed.). *Forms of Knowledge in Early Modern Asia: Explorations in the Intellectual History of India and Tibet, 1500–1800.* Duke University Press, Durham, NC. pp.: 291–310.

Senda M. 1992. *Japan's Traditional View of Nature and Interpretation of Landscape*, /in:/ GeoJournal. Vol. 26. No. 2. pp.: 129–134.

Sharma T. 2018. *In search of the truth in the sacred art of India: perspectives on authenticity* /in:/ Wijesuriya G., Sweet J. (eds.). 2018. *Revisiting Authenticity in the Asian Context.* Conservation Forum Series. ICCROM, Rome. Vol. 2. pp.: 171–180.

Silva R. 1983. *The Significance of the Venice International Charter for the Conservation and Restoration of Monuments and Sites, with Special Reference to Eastern Countries.* Institute for International Art Festivals, ICOMOS & ICCROM. (eds.). First International Congress on Architectural Conservation: Historic Buildings, their Significance and their Role in Today's Cultural Setting. University of Basle, March 1983. pp.: 10–14. http://www.icomos.org/venice charter2004/silva.pdf. (accessed: 07. 11. 2018).

Steinhardt S. N. (ed.). 2002. *Chinese Architecture*. Yale University, New Haven.
Steinhardt S. N. 1994. *Liao: An Architectural Tradition in the Making*. /in:/ Artibus Asiae, no. 54. pp.: 5–39.
Steinhardt S. N. 1997. *Liao architecture*. University of Hawaii Press, Honolulu.
Steinhardt S. N. 2004. *The Tang Architectural Icon and the Politics of Chinese Architectural History*. /in:/ The Art Bulletin. Vol. 86. No. 2, 2004. pp.: 228–254.
Stovel H. 2004. *Authenticity in conservation decision-making: the World Heritage perspective*. / in:/ Journal of Research in Architecture and Planning, Conservation and Cultural Heritage. Vol.: 3. pp.: 1–8.
Stovel H. 2007. *Effective use of authenticity and integrity as World Heritage qualifying conditions*. /in:/ City & Time. Vol. 2, no. 3. pp.: 21–36.
Technical code for maintenance and strengthening of ancient timber buildings. 1993. National Technical Committee 41 on Timber of Standardization Administrator of China, GB 50165-92. China Building Industry Press, Beijing.
The Athens Charter for the Restoration of Historic Monuments – 1931. 2017. Congress of Architects. Paris. http.//www.icomos.org/en/charters-and-texts/179-articles-enfrancais/ressources/charters-and-standards/167-the-athenscharter-for-the-restoration-of-historic-monuments. (accessed: 20.10. 2018).
The Law of the People's Republic of China on Protection of Cultural Relics. 2007. Selection of International Charters and National Regulations on the Protection of Urban Cultural Heritage Zhang S. (ed.), Tongji University Press, Shanghai pp.: 157–166 (in Chinese).
Tiwari S.R. 2009. *Material Authenticity in Tradition of Conservation of Nepal*. Paper presented at the Thunder International Conference, October 2009, Kathmandu. Kathmandu. http://www.kailashkut.com/wp-content/uploads/2016/05/materialauthenticity.pdf. (accessed: 10. 03. 2018).
Tiwari S.R. 2015. *Revisiting Kathmandu Valley World Heritage site: Community participation in heritage affairs*. /in:/ Weise K. (ed.). 2015. *Revisiting Kathmandu Safeguarding Living Urban Heritage*. Proceedings of an international symposium Kathmandu Valley, 25–29 November 2013. Paris and Kathmandu, UNESCO and UNESCO Office in Kathmandu. pp.: 189–196.
UNESCO. 1972. *Convention concerning the Protection of the World Cultural and Natural Heritage. (World Heritage Convention)*. Paris. http://whc.unesco.org/en/conventiontext. (accessed: 03. 09. 2018).
UNESCO. 2001. *UNESCO Universal Declaration on Cultural Diversity*. Paris. https://adsdatabase.ohchr.org/IssueLibrary/UNESCO%20Universal%20Declaration%20on%20Cultural%20Diversity.pdf. (accessed: 13. 05. 2019).
UNESCO. 2009. *Foguang Temple Nomination File*. https://whc.unesco.org/en/list/1279/documents/. (accessed: 02. 03. 2021)
UNESCO. 2016. *Operational Guidelines for the Implementation of the World Heritage Convention*. World Heritage Centre, Paris. http://whc.unesco.org/en/guidelines. (accessed: 10. 06. 2019).
UNESCO. 2017. *The 2005 Convention on the Protection and Promotion of the Diversity of Cultural Expressions*. Paris. http://en.unesco.org/creativity/convention. (accessed: 15. 08. 2019).
UNESCO. 2017. *The Convention for the Safeguarding of Intangible Cultural Heritage*. Santiago. http://www.unesco.org/new/en/santiago/culture/intangible-heritage/conventionintangible-cultural-heritage. (accessed: 13. 09. 2019).
UNESCO. Undated. *Old Town of Lijiang*. http://whc.unesco.org/en/list/811. (accessed October 20, 2020).
Varga S., Guignon C. 2017. *Authenticity*. /in:/ Zalta E.N. (ed.). The Stanford Encyclopedia of Philosophy. Stanford. https://plato.stanford.edu/entries/authenticity. (accessed: 11. 12. 2021).
Wang F. 2016. *Geo-Architecture and Landscape in China's Geographic and Historic Context*. Springer.
Wang J. 2009. *Authenticity*. /in:/ Urban Planning, vol. 11 (in Chinese).

Weerasinghe J. 2006. *Living sacred sites and the concept of authenticity.* /in:/ Proceedings of the 14th Seminar on the Conservation of Asian Cultural Heritage: Living with Cultural Heritage. National Research Institute for Cultural Properties, Tokyo. pp.: 179–188.

Weerasinghe J. 2011. *Living Sacred Heritage and 'Authenticity' in South Asia.* /in:/ Anheier H., Isar Y.R. (eds.). *Heritage, Memory and Identity.* SAGE Publications Ltd., London. pp.: 139–147.

Weiler H.N. 2017. *Whose Knowledge Matters? Development and the Politics of Knowledge.* Stanford. http://web.stanford.edu/~weiler/Texts09/Weiler_Molt_09.pdf. (accessed: 10. 02. 2021).

Weise K. (ed.). 2015. *Revisiting Kathmandu Safeguarding Living Urban Heritage.* Proceedings of an international symposium Kathmandu Valley, 25–29 November 2013. Paris and Kathmandu, UNESCO and UNESCO Office in Kathmandu.

Weise K. 2015. *Summary and commentary.* /in:/ Weise K. (ed.). *Revisiting Kathmandu: Safeguarding Living Urban Heritage.* UNESCO, Paris. pp.: 1–47.

Weise K. 2018. *Authenticity and the safeguarding of cultural heritage in Nepal* /in:/ Wijesuriya G., Sweet J. (eds.). 2018. *Revisiting Authenticity in the Asian Context.* Conservation Forum Series. ICCROM, Rome. Vol. 2. pp.: 119–132.

Wijesuriya G. 2005. *The past is in the present: Perspectives in Caring for Buddhist Heritage* /in:/ Stovel H., Stanley-Price N., Killick R. (eds.). 2005. *Conservation of Living Religious Heritage.* Papers from the ICCROM 2003 Forum on Living Religious Heritage: "Conserving the Sacred". ICCROM Conservation Studies 3. ICCROM, Rome. pp.: 31–43.

Wijesuriya G. 2007. *Conserving Living Taonga: The Concept of Continuity.* /in:/ Sully D. (ed.). *Decolonising Conservation – Caring for Maori Meeting Houses outside New Zealand.* Left Coast Press, Walnut Creek. pp.: 59–69.

Wijesuriya G. 2008. *Values of the Heritage in the Religious and Cultural Traditions of Southern Asia.* /in:/ Bouchenaki M., Tomaszewski A., del Bianco P. (eds.). *Values and Criteria in Heritage Conservation,* Proceedings of the International Conference, 2–4 March 2007, Edizioni Polistampa, Firenze. pp.: 73–78.

Wijesuriya G. 2010. *Conservation in Context.* /in/ Falser M.S., Lipp W., Tomaszewski A. (eds.). *Proceedings of the International Conference on 'Conservation and Preservation- Interaction between Theory and Practice, In memoriam Alois* Riegl *(1858-1905)'.* Edizioni Polistampa, Firenze. pp.: 233–248.

Wijesuriya G. 2017. *Towards De-secularisation of Heritage.* /in:/ Built Heritage. Vol.: 1(2). pp.: 1–15.

Wijesuriya G. 2018. *Living Heritage.* /in:/ Heritage A., Copithorne J. (eds.). *Sharing Conservation Decisions: Current Issues and Future Strategies.* Published by the International Centre for the Study of the Preservation and Restoration of Cultural Property (ICCROM), Rome. pp.: 43–56.

Wijesuriya G., Sweet J. (eds.). 2018. *Revisiting Authenticity in the Asian Context.* Conservation Forum Series. ICCROM, Rome. Vol. 2.

Wilkinson E. P. 2000. *Chinese History: A Manual. Revised and Enlarged.* Harvard University, Harvard University Asia Center, Cambridge, Mass.

Xia R. 2013. *"Noah's Ark" in Nanchan Temple's Dilemma.* /in:/ Global Human Geography. Southwest University, 21st Century Talent News, Chongqing City. (in Chinese).

Xiannian F. 2001. *Research on the layout and architectural design methods of ancient Chinese urban planning buildings.* China Architecture Industry Press, Beijing. (in Chinese).

Xie Q.F. 2007. *Experimental study and theoretical analysis on strengthening for Chinese ancient timber buildings.* College of Civil Engineering, Xi'an University of Architecture & Technology, Xi'an.

Xu S. 2005. *Cultural and Natural Heritage Conservation in China.* Science Publisher, Beijing (in Chinese).

Yao D. 2013. *Discovering the Dule Temple,* /in:/ Architectural Journal. Issue 5/2013. (in Chinese).

Ye H., Dai X. 2002. *A study on the restoration of Hu Xueyan's former residence: challenges for the Venice Charter.* /in:/ Guangming Daily, 2 October 2002 (in Chinesse).

Yih-yuan L. 1979. *Chinese Geomancy and Ancestor Worship: A Further Discussion, Ancestors.* Newell W. (ed.). The Hague, Mouton pp.: 329–338.

Yu Z.M. 2014. *Structure chart analysis of Yingxian Wooden Pagoda* (dissertation). Tianjin University, Tianjin (in Chinese).

Ze-li Q., Zhe-rui L., Xiao-lan Z., Zi-ye Y. and Biao P. 2017. *Traditional Wooden Buildings in China.* /in:/ Concu G. (ed.) *Wood in Civil Engineering.* Intech Open Limited, London, United Kingdom. pp. 197–221.

Zerrudo E. B. 2018. *If walls could speak: authenticity in the Philippine context* /in:/ Wijesuriya G., Sweet J. (eds.). 2018. *Revisiting Authenticity in the Asian Context.* Conservation Forum Series. ICCROM, Rome. Vol. 2. pp.: 103–117.

Zerrudo E.B. 2008. *Pamanaraan: Writings on Philippine Heritage Management.* University of Santo Tomas Publishing House, Manila.

Zhan G. 2018. *Discussion on authenticity in China: reflections on the Venice Charter.* /in:/ Wijesuriya G., Sweet J. (eds.). 2018. *Revisiting Authenticity in the Asian Context.* Conservation Forum Series. ICCROM, Rome. Vol. 2. pp.: 89–94.

Zhang C. 2010. *The Discussion about Two Translations of Authenticity.* /in:/ Journal of Architecture, Vol. 2. pp.: 55–59 (in Chinese).

Zhang F.L. 2013. *Research on strengthening and its performance of Chinese ancient timber structures* (dissertation). Xi'an University of Architecture & Technology; Xi'an (in Chinese).

Zhang J. 2007. *Authenticity in Heritage Conservation Institution in Old City: Truth or Error.* /in:/ City Planning Review, 11/20. pp.: 79–83 (in Chinese).

Zhang R. 2010. *Survey of the east palace of Foguang Temple.* /in:/ Traditional Chinese Architecture and Gardens, Volume 3, pp.: 29–39 (in Chinese).

Zhao L. 2010. *World Treasure Nanchan Temple.* Wutai Mountain Research. Wutai Mountain Research Association, Taiyuan City. (in Chinese).

Zheng J. 2014. *Comparison of Heritage Conservation Philosophies in China and Other Countries.* / in:/ International Principles and Local Practices of Cultural Heritage Conservation Conference Proceedings. Tsinghua University/ ICOMOS China, Beijing. pp.: 40–61.

Zhou L. 2018. *Authenticity & the Principles for the Conservation of Heritage Sites in China* /in:/ Wijesuriya G., Sweet J. (eds.). 2018. *Revisiting Authenticity in the Asian Context.* Conservation Forum Series. ICCROM, Rome. Vol. 2. pp.: 95–101.

Zhu G. 2012. *China's architectural heritage conservation movement.* /in:/ Frontiers of Architectural Research, no. 1. pp.: 10–22.

Zhu Y. 2015. *Cultural Effects of Authenticity: Contested Heritage Practices in China.* /in:/ International Journal of Heritage Studies 21/6. pp.: 594–608.

Zhu Y. 2017. *Authenticity and Heritage Conservation in China: Translation, Interpretation and Practices* /in:/ Authenticity in Architectural Heritage Conservation; Transcultural Research – Heidelberg Studies on Asia and Europe in a Global Context. Weiler K., Gutschow N. (eds.). Springer International Publishing, Switzerland. pp.: 187–200.

Zi Y. 2012. *Famous Temples in China.* Huangshan Publishing House, Hefei, Anhui. pp.: 57–60. (in English and Chinese).

Zuyi J. 1992. *The whole Liao Dynasty poetry.* Yuelu Publishing House, Beijing. (in Chinese).

FSC
www.fsc.org
MIX
Papier | Fördert gute Waldnutzung
FSC® C083411

Kolibri 360 GmbH
Ferdinand-Jühlke-Straße 7
99095 Erfurt, Deutschland
produktsicherheit@kolibri360.de